Student's Solutions Manual to Accompany

SECOND EDITION

Intermediate Algebra

Student's Solutions Manual to Accompany

SECOND EDITION

Intermediate Algebra

Daniel L. Auvil

KENT STATE UNIVERSITY

▲▼ ADDISON-WESLEY PUBLISHING COMPANY

Reading, Massachusetts   Menlo Park, California
Don Mills, Ontario   Wokingham, England   Amsterdam
Sydney   Singapore   Tokyo   Madrid   Bogotá
Santiago   San Juan

ISBN 0-201-11047-4
ABCDEFGHIJ-AL-8987

CONTENTS

Preface

This supplement is designed to aid those students who are studying mathematics from the text <u>Intermediate Algebra</u> (second edition) by Daniel L. Auvil. It contains step-by-step solutions to all the even-numbered problems in the text, except for those in the Chapter Reviews, Chapter Tests, and Cumulative Reviews.

Besides providing detailed solutions to the even-numbered problems, the supplement can be helpful in solving the odd-numbered problems. Since the problems in the text are generally matched so that in any given problem set Problem 1 is similar to Problem 2, Problem 3 is similar to Problem 4, and so on, the solution to a particular odd-numbered problem can usually be determined by noting the procedure for solving the corresponding even-numbered problem. This is not necessarily true for the calculator problems or the challenging problems, however.

Instructors can use the supplement to their advantage by assigning even-numbered problems for homework. This reduces the amount of class time spent answering questions on homework problems, thereby allowing more time for introducing new topics in a more thorough and enjoyable fashion.

Many thanks to Debbie Southworth who accomplished the difficult task of typing this camera-ready supplement, and to Gloria E. Langer who did an excellent job proofreading the final copy.

N. Canton, Ohio

D. L. A.

CHAPTER 1

REAL NUMBERS AND THEIR PROPERTIES

Problem Set 1.1, pp. 7-9

2. $x - 9$

4. $2 \cdot y$

6. $\frac{2}{3} \cdot z$

8. $10 \cdot t$

10. $s + 6$

12. $1.95r$

14. $2(m - 5)$

16. $379x + 11.80y$

18. $127 + 3 \cdot 5$

   $= 127 + 15$

   $= 142$ dollars

20. $28 - 8 \cdot 2 + 9$

   $= 28 - 16 + 9$

   $= 12 + 9$

   $= 21$

22. $5 + 6 \cdot 7 - 4$

   $= 5 + 42 - 4$

   $= 47 - 4$

   $= 43$

24. $3 + 12 \div 3 + 3$

   $= 3 + 4 + 3$

   $= 10$

26. $3 + 12 \div (3 + 3)$

   $= 3 + 12 \div 6$

   $= 3 + 2$

   $= 5$

28. $28 \div 4 - 2$

   $= 7 - 2$

   $= 5$

30. $16 \div 8 \cdot 2$

   $= 2 \cdot 2$

   $= 4$

32. $\frac{4}{5} \cdot \frac{1}{3} + \frac{14}{15}$

   $= \frac{4}{15} + \frac{14}{15}$

   $= \frac{18}{15}$

   $= \frac{6}{5}$

34. $\frac{2}{3} \cdot (\frac{5}{2} - \frac{7}{10})$

$= \frac{2}{3}(\frac{25}{10} - \frac{7}{10})$

$= \frac{2}{3}(\frac{18}{10})$

$= \frac{\overset{1}{\cancel{2}}}{\underset{1}{\cancel{3}}} \cdot \frac{\overset{6}{\cancel{18}}}{\underset{5}{\cancel{10}}}$

$= \frac{6}{5}$

36. $\frac{8}{9} + \frac{8}{15} \div 3\frac{1}{5}$

$= \frac{8}{9} + \frac{8}{15} \div \frac{16}{5}$

$= \frac{8}{9} + \frac{\overset{1}{\cancel{8}}}{\underset{3}{\cancel{15}}} \cdot \frac{\overset{1}{\cancel{5}}}{\underset{2}{\cancel{16}}}$

$= \frac{8}{9} + \frac{1}{6}$

$= \frac{16}{18} + \frac{3}{18}$

$= \frac{19}{18}$

38. $6[9 - (2 + 7)]$

$= 6[9 - 9]$

$= 6[0]$

$= 0$

40. $5[14 - 4(3 - 1)]$

$= 5[14 - 4(2)]$

$= 5[14 - 8]$

$= 5[6]$

$= 30$

42. $30[(\frac{1}{2} - \frac{2}{5}) + (\frac{5}{9} + \frac{4}{9})]$

$= 30[(\frac{5}{10} - \frac{4}{10}) + (\frac{9}{9})]$

$= 30[\frac{1}{10} + 1]$

$= 30[\frac{1}{10} + \frac{10}{10}]$

$= \overset{3}{\cancel{30}}[\frac{11}{\underset{1}{\cancel{10}}}]$

$= 33$

44. $8 + [4 + 2(3 + 6)]$

$= 8 + [4 + 2(9)]$

$= 8 + [4 + 18]$

$= 8 + 22$

$= 30$

46. $3(2[12 - 2(6 - 5)])$

    $= 3(2[12 - 2(1)])$

    $= 3(2[12 - 2])$

    $= 3(2[10])$

    $= 60$

48. $81 \div [14 - 4(10 - 8)]$

    $= 81 \div [14 - 4(2)]$

    $= 81 \div [14 - 8]$

    $= 81 \div 6$

    $= 13.5$

50. $89.09 - [3.4(9.6 + 2.8) - 7.5]$

    $= 89.09 - [3.4(12.4) - 7.5]$

    $= 89.09 - [42.16 - 7.5]$

    $= 89.09 - 34.66$

    $= 54.43$

52. $\dfrac{36 - 10}{4 + 2} \cdot \dfrac{12 + 8}{4 \cdot 3}$

    $= \dfrac{\overset{13}{\cancel{26}}}{\underset{3}{\cancel{6}}} \cdot \dfrac{\overset{5}{\cancel{20}}}{\underset{3}{\cancel{12}}}$

    $= \dfrac{65}{9}$

54. $\dfrac{17 - 4(\frac{8 + 6}{7}) + 13}{38 + 38 \div 2 - 19}$

    $= \dfrac{17 - 4(2) + 13}{38 + 19 - 19}$

    $= \dfrac{17 - 8 + 13}{38}$

    $= \dfrac{22}{38}$

    $= \dfrac{11}{19}$

56. $4x - 6y + z$

    $= 4(8) - 6(3) + 2$

    $= 32 - 18 + 2$

    $= 14 + 2$

    $= 16$

58. $x - yz$

    $= 8 - 3 \cdot 2$

    $= 8 - 6$

    $= 2$

60. $2(x + y)$

    $= 2(8 + 3)$

    $= 2(11)$

    $= 22$

62. $(3\frac{1}{4} + 4\frac{3}{4} + 5\frac{3}{4}) \div 3$

    $= 12\frac{7}{4} \div 3$

    $= \dfrac{55}{4} \cdot \dfrac{1}{3}$

    $= \dfrac{55}{12}$ or $4\frac{7}{12}$ lb

64. $28 \cdot 1\frac{3}{8}$

$= \frac{\overset{7}{\cancel{28}}}{1} \cdot \frac{11}{\underset{2}{\cancel{8}}}$

$= \frac{77}{2}$

$= 38\frac{1}{2}$ in.

66. $24 - (7\frac{3}{4} + 3\frac{2}{3})$

$= 24 - (7\frac{9}{12} + 3\frac{8}{12})$

$= 24 - 10\frac{17}{12}$

$= 23\frac{12}{12} - 11\frac{5}{12}$

$= 12\frac{7}{12}$ in.

68. $28(15.50) + 28(8.50) = 434 + 238 = \$672$

$28(15.50 + 8.50) = 28(24) = \$672$

70. $7(x + 2)$

$= 7 \cdot x + 7 \cdot 2$

$= 7x + 14$

72. $4(3y + 8)$

$= 4 \cdot 3y + 4 \cdot 8$

$= 12y + 32$

74. $15(\frac{1}{3}z + \frac{3}{5})$

$= 15 \cdot \frac{1}{3}z + 15 \cdot \frac{3}{5}$

$= 5z + 9$

76. $3(a + b + c)$

$= 3 \cdot a + 3 \cdot b + 3 \cdot c$

$= 3a + 3b + 3c$

78. $36(\frac{v}{4} + \frac{2}{9})$

$= 36 \cdot \frac{v}{4} + 36 \cdot \frac{2}{9}$

$= 9v + 8$

80. $7b + 7c$

$= 7 \cdot b + 7 \cdot c$

$= 7(b + c)$

82. $3 \cdot x + 3 \cdot 5$

$= 3(x + 5)$

84. $6x + 24$

$= 6 \cdot x + 6 \cdot 4$

$= 6(x + 4)$

86. $11r + 11s + 11t + 11u$

$= 11(r + s + t + u)$

88. $2[3(x + 7) - 9] \div 6 - x$

$= 2[3x + 21 - 9] \div 6 - x$

$= 2[3x + 12] \div 6 - x$

$= (6x + 24) \div 6 - x$

$= x + 4 - x$

$= 4$

90. $\dfrac{a}{b} - \dfrac{c}{d} = \dfrac{ad}{bd} - \dfrac{bc}{bd} = \dfrac{ad - bc}{bd}$

92. a) $1712(1978 + 51,327) = 1712(53,305) = 91,258,160$

    b) $1712(1978 + 51,327) = 1712(1978) + 1712(51,327)$

$$= 3,386,336 + 87,871,824$$

$$= 91,258,160$$

Problem Set 1.2, pp. 15–17

2. -10 sec, $+3$ sec

6. $+23°$, $-6°$

4. $+1700$ ft, -355 ft

8.
$$4 \overline{)1.00} \quad \begin{array}{r} 0.25 \\ \hline \end{array}$$

```
      0.25
  4 )1.00
      8
      20
      20
       0
```

$\dfrac{1}{4} = 0.25 = 25\%$

10.
```
      0.8
  5 )4.0
      4 0
        0
```

$7\dfrac{4}{5} = 7.8 = 780\%$

12.
```
       0.875
  8 )7.000
      6 4
        60
        56
        40
        40
         0
```

$\dfrac{7}{8} = 0.875 = 87.5\%$

14.
$$\begin{array}{r} 0.44\overline{4} \\ 9\overline{)4.000} \\ \underline{3\ 6} \\ 40 \\ \underline{36} \\ 40 \\ \underline{36} \\ 4 \end{array}$$

$\frac{4}{9} = 0.44\overline{4} = 44.\overline{4}\%$

16.
$$\begin{array}{r} 0.18\overline{18} \\ 11\overline{)2.0000} \\ \underline{1\ 1} \\ 90 \\ \underline{88} \\ 20 \\ \underline{11} \\ 90 \\ \underline{88} \\ 2 \end{array}$$

$\frac{2}{11} = 0.18\overline{18} = 18.\overline{18}\%$

18.
$$\begin{array}{r} 0.59\overline{09} \\ 22\overline{)13.0000} \\ \underline{11\ 0} \\ 2\ 00 \\ \underline{1\ 98} \\ 20 \\ \underline{0} \\ 200 \\ \underline{198} \\ 2 \end{array}$$

$\frac{13}{22} = 0.59\overline{09} = 59.\overline{09}\%$

20.
$$\begin{array}{r} 6.5 \\ 2\overline{)13.0} \\ \underline{12} \\ 1\ 0 \\ \underline{1\ 0} \\ 0 \end{array}$$

$\frac{-13}{2} = -6.5 = -650\%$

22. $\frac{0}{5} = 0 = 0\%$

24. $\frac{5}{0}$ is undefined.

26. a) 1, 119

b) 1, 119, −0

c) 1, 119, −0, −100

d) All of them

28. Smallest to largest:

$1.08 = 1.0800$

$1.\overline{08} = 1.08\overline{08}$

$\frac{9}{5} = 1.8000$

$1.\overline{8} = 1.8888.$

30. 6.75, for example.

Infinite number

34. $9.99\overline{9} = 3(3.33\overline{3})$

$= 3(\frac{10}{3})$

$= 10$

32. a) $3.1 = 3\frac{1}{10} = \frac{31}{10}$

b) $0.00239 = \frac{239}{100,000}$

36. a) $\sqrt{36}, -\frac{22}{7}, 6.17, 0.\overline{05}$

b) $-\pi$, $0.8181181118..., \sqrt{18}$

38. False

42. False

46. True

50. $5 \cdot (-3) = (-3) \cdot 5$

54. $(x + 3) + (-3) = x + (3 + (-3))$

56. $5 \cdot (8 \cdot \frac{1}{2}) = (5 \cdot 8) \cdot \frac{1}{2}$

60. $19 + 0 = 19$

64. $(-5) + 5 = 0$

68. $(-3)[9 + (-2)] = (-3)9 + (-3)(-2)$

70. $5x + 6x = (5 + 6)x$

74. $12 \div (6 \div 2) \neq (12 \div 6) \div 2$

78. $m + 2$

82. 0.0625s

40. True

44. True

48. $9 + 4 = 4 + 9$

52. $1 + (7 + 9) = (1 + 7) + 9$

58. $\frac{1}{4} \cdot (4x) = (\frac{1}{4} \cdot 4)x$

62. $1 \cdot 8 = 8$

66. $7 \cdot \frac{1}{7} = 1$

72. $3 - 2 \neq 2 - 3$

76. $n - 1$

80. 0.57r

84. 0.30t

86. a) Since 2n is even, 2n + 2 is even.

b) Since 2n is even, 2n + 3 is odd.

88. The probability that no two have the same birthday is

$$\frac{365}{365} \cdot \frac{364}{365} \cdot \frac{363}{365} \cdots \cdot \frac{366-40}{365}$$

$$= \frac{365}{365} \cdot \frac{364}{365} \cdot \frac{363}{365} \cdots \cdot \frac{326}{365}$$

$$\approx 1(0.9972603)(0.9945206) \cdots \cdot (0.8931507)$$

$$\approx 0.109$$

$$\approx 10.9\%$$

Therefore the probability that at least two persons will have the same birthday is approximately

$$100\% - 10.9\% = 89.1\%$$

90. a) $\frac{2}{3} + \frac{4}{11} = 0.6\overline{6} + 0.\overline{36} = 1.\overline{03}$

    b) $\frac{2}{11} + \frac{4}{27} = 0.\overline{18} + 0.\overline{148} = 0.\overline{329966}$

    c) $\frac{2}{7} + \frac{3}{7} = 0.\overline{285714} + 0.\overline{428571} = 0.\overline{714285}$

Problem Set 1.3, pp. 22-24

2.

4. $-7 < -4$ 6. $-100 < -99$

8. $0 > -9$ 10. $-7 < 4$

12. Since $\pi \approx 3.1416$, $\pi > 3.14$.

14. Since $-\sqrt{3} \approx -1.^{-}32$, $-\sqrt{3} < -1.7$.

16. $\frac{4}{5} = \frac{36}{45}$ 18. $-\frac{4}{5} = -\frac{36}{45}$

    $\frac{7}{9} = \frac{35}{45}$ $-\frac{7}{9} = -\frac{35}{45}$

    So $\frac{4}{5} > \frac{7}{9}$.   So $-\frac{4}{5} < -\frac{7}{9}$.

20. $x < 0$ 22. $y \le 0$

24. $z \ge 4$ 26. $r - 3 < 17$

28. $-1 < s < 0$ 30. $6 < t \le 23$

32. $d \ge 19$ yr   34. 800 ft/sec $< s <$ 4000 ft/sec

36. If Ted lives on a line segment drawn from Jerry's house to the
     school, then Jerry lives 5 + 3 = 8 miles from the school. If
     Jerry lives on a line segment drawn from Ted's house to the
     school, then Jerry lives 5 - 3 = 2 miles from the school. This
     means that Jerry lives anywhere from 2 miles to 8 miles from
     school. Therefore, the desired double inequality is
     2 mi $\le d \le$ 8 mi.

38. a) Minimum price = cost + 60% markup

                       = 18.70 + 0.60(18.70)

                       = 18.70 + 11.22

                       = 29.92

     Therefore $29.92 $\le p <$ $32.

     b) Can't be done since the minimum price is $29.92.

40. 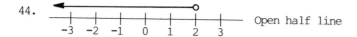 Three points

42.  Infinite number of points

44.  Open half line

46. 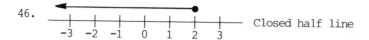 Closed half line

48. 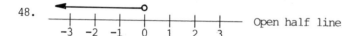 Open half line

50. 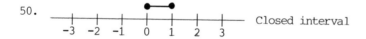 Closed interval

52. 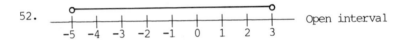   Open interval

54. Half-open interval

56. Half-open interval

58. $\left\{ x \mid x < 2 \text{ and } x \text{ is an integer} \right\}$

60. $\left\{ x \mid x \leq 1 \right\}$

62. $\left\{ x \mid -3 < x \leq 2 \right\}$

64. $|14| = 14$

66. $|-8| = 8$

68. $\left| -4\frac{2}{3} \right| = 4\frac{2}{3}$

70. $\left| -\sqrt{5} \right| = \sqrt{5}$

72. $|-0| = 0$

74. $|\pi| = \pi$

76. $-|35| = -35$

78. $-|-101| = -101$

80. $\dfrac{|-6| + |-9|}{|-3|} = \dfrac{6 + 9}{3} = \dfrac{15}{3} = 5$

82. $\dfrac{15}{|6| - |-6|} = \dfrac{15}{6 - 6} = \dfrac{15}{0}$, which is undefined.

84. $\left| \dfrac{18 - 6(2 + 1)}{9} \right| = \left| \dfrac{18 - 6(3)}{9} \right| = \left| \dfrac{18 - 18}{9} \right| = \left| \dfrac{0}{9} \right| = |0| = 0$

86. $\left| \dfrac{20 - 15}{10 + 5} \right| = \left| \dfrac{5}{15} \right| = \left| \dfrac{1}{3} \right| = \dfrac{1}{3}$

88. True

90. False. For example, $|-3| > |2|$, but $-3 \ngtr 2$.

92. Have each occupant of the Royal double their room number and move to that room. This will open up all the odd-numbered rooms.

94. $\dfrac{68,847}{39.75} \leq n \leq \dfrac{77,115}{39.75}$

   1732 sq ft $\leq n \leq$ 1940 sq ft

Problem Set 1.4, pp. 29-31

2. $(-4) + (-3) = -7$

4. $4 + (-7) = -3$

6. $16 + (-2) = 14$

8. $(+8) + (+9) = 17$

10. $(-15) + 0 = -15$

12. $(-4) + (-12) = -16$

14. $(-5.3) + 2.8 = -2.5$

16. $-19.1 + 12.36 = -6.74$

18. $(-\frac{13}{7}) + (-\frac{2}{7}) = -\frac{15}{7}$

20. $\frac{3}{4} + (-\frac{1}{6}) = \frac{9}{12} + (-\frac{2}{12}) = \frac{7}{12}$

22. $-896 + (-547) = -1443$

24. $\quad 14.40 + (-6) + (-4) + 5.80 + 9.60 + (-23.20)$

$= [14.40 + 5.80 + 9.60] + [-6 + (-4) + (-23.20)]$

$= 29.80 + (-33.20)$

$= \$-3.40$

26. $\frac{-3 + 15 + (-1) + 3 + (-4)}{5} = \frac{10}{5} = 2$ yd per carry

28. $-a = -17$

30. $-a = -(-6) = 6$

32. $5 - 3 = 5 + (-3) = 2$

34. $3 - 5 = 3 + (-5) = -2$

36. $(-8) - 5 = -8 + (-5) = -13$

38. $(-15) - (-4) = -15 + 4 = -11$

40. $\frac{2}{9} - \frac{8}{9} = \frac{2}{9} + (-\frac{8}{9}) = -\frac{6}{9} = -\frac{2}{3}$

42. $0 - 8 = 0 + (-8) = -8$

44. $(-9) - (-22) = -9 + 22 = 13$

46. $(-\frac{5}{6}) - (-\frac{3}{5}) = -\frac{5}{6} + \frac{3}{5} = -\frac{25}{30} + \frac{18}{30} = -\frac{7}{30}$

48. $0 - (-\frac{1}{2}) = 0 + \frac{1}{2} = \frac{1}{2}$

50. $4.9 - 9.21 = 4.9 + (-9.21) = -4.31$

52. $-8.7 - 5.8 = -8.7 + (-5.8) = -14.5$

54. $-712 - 629 = -712 + (-629) = -1341$

56. $-13 - 28.67 = -13 + (-28.67) = \$-41.67$

58. $10 - 3 \cdot 7 = 10 - 21 = 10 + (-21) = -11°$

60. $-7 + 13 + 5$

$= 6 + 5$

$= 11$

62. $2 - (-24) - 6$

$= 2 + 24 - 6$

$= 26 - 6$

$= 20$

64. $-11 - (6 - 4) + (-18 + 7)$

$= -11 - 2 + (-11)$

$= -13 + (-11)$

$= -24$

66. $-\dfrac{2}{3} - (\dfrac{1}{2} - \dfrac{7}{9})$

$= -\dfrac{2}{3} - (\dfrac{9}{18} - \dfrac{14}{18})$

$= -\dfrac{2}{3} - (-\dfrac{5}{18})$

$= -\dfrac{12}{18} + \dfrac{5}{18}$

$= -\dfrac{7}{18}$

68. $\left| \dfrac{7 - 2(5 + 1)}{29 - 10} \right|$

$= \left| \dfrac{7 - 2(6)}{19} \right|$

$= \left| \dfrac{7 - 12}{19} \right|$

$= \left| \dfrac{-5}{19} \right|$

$= \dfrac{5}{19}$

70. $|-16| - |-5| - (|-8| - |3|)$

$= 16 - 5 - (8 - 3)$

$= 16 - 5 - 5$

$= 11 - 5$

$= 6$

72. $54 - 28 - (16 - 30) - [63 - (-5 - 8 + 4)]$

$= 54 - 28 - (-14) - [63 - (-13 + 4)]$

$= 54 - 28 + 14 - [63 - (-9)]$

$= 54 - 28 + 14 - [63 + 9]$

$= 54 - 28 + 14 - 72$

$= 26 + 14 - 72$

$= 40 - 72$

$= -32$

74. $r + t = 5 + (-9) = -4$

76. $r + s - t = 5 + (-8) - (-9) = 5 + (-8) + 9 = 6$

78. $r - (s - t) + t = 5 - (-8 - (-9)) + (-9)$

$$= 5 - (-8 + 9) + (-9)$$

$$= 5 - 1 + (-9)$$

$$= 4 + (-9)$$

$$= -5$$

80. Let $a = -5$ and $b = 3$. Then $|a - b| = |-5 - 3| = |-8| = 8$, but $|-5| - |3| = 5 - 3 = 2$.

82.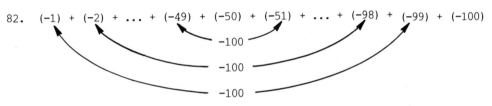

$(-1) + (-2) + \ldots + (-49) + (-50) + (-51) + \ldots + (-98) + (-99) + (-100)$

$$= 50(-100) + (-50)$$

$$= -5000 + (-50)$$

$$= -5050$$

Problem Set 1.5, pp. 34-36

2. $9(-3) = -27$

4. $(-4)6 = -24$

6. $(-2)(-8) = 16$

8. $19(-1) = -19$

10. $44 \cdot 0 \cdot (-19) = 0$

12. $4 \cdot 5 \cdot (-10) = 20(-10) = -200$

14. $\frac{1}{3}(-\frac{1}{5}) = -\frac{1}{15}$

16. $(-\frac{1}{2})(-\frac{1}{2}) = \frac{1}{4}$

18. $(-4)(-4)(-4) = 16(-4) = -64$

20. $(-2)(-2)(-2)(-2) = 4 \cdot 4 = 16$

22. $91.6(-8.15) = -746.54$

24. $16(-0.50) + 2 \cdot 5 + 3 \cdot 4$

$$= -8 + 10 + 12$$

$$= \$14$$

26. $(-15) \div (-5) = 3$

28. $(-35) \div (-7) = 5$

30. $\dfrac{-20}{5} = -4$

32. $\dfrac{-17}{17} = -1$

34. $\dfrac{18}{-3} = -6$

36. $\dfrac{0}{-12} = 0$

38. $\dfrac{-240}{-20} = 12$

40. $\dfrac{46.72}{-6.4} = -7.3$

42. $(-\dfrac{3}{4}) \div \dfrac{2}{5} = -\dfrac{3}{4} \cdot \dfrac{5}{2} = -\dfrac{15}{8}$

44. $(-\dfrac{5}{8}) \div (-\dfrac{25}{24}) = -\overset{1}{\underset{1}{\dfrac{5}{8}}} \cdot (-\overset{3}{\underset{5}{\dfrac{24}{25}}}) = \dfrac{3}{5}$

46. $\dfrac{-27}{0}$ is undefined.

48. $\dfrac{2 + (-4) + (-6) + (-7) + (-10) + 0 + 4}{7} = \dfrac{-21}{7} = -3°$

50. $-\dfrac{-3}{5} = \dfrac{-3}{-5} = \dfrac{3}{5}$

52. $-\dfrac{16}{8} = -2$

54. $\dfrac{-16}{-8} = 2$

56. $-\dfrac{1}{-4} = \dfrac{-1}{-4} = \dfrac{1}{4}$

58. $-\dfrac{-x}{-19} = -\dfrac{x}{19}$

60. $\dfrac{1}{a} = \dfrac{1}{8}$

62. $\dfrac{1}{a} = \dfrac{1}{1/8} = 8$

64. $\dfrac{1}{a} = \dfrac{1}{5/6} = \dfrac{6}{5}$

66. $\dfrac{1}{a} = \dfrac{1}{-26/5} = -\dfrac{5}{26}$

68. $\dfrac{1}{a} = \dfrac{1}{-1} = -1$

70. $\dfrac{1}{a} = \dfrac{1}{p/q} = \dfrac{q}{p}$

72. $(-8) \cdot 9 + 3 \cdot 6$

      $= -72 + 18$

      $= -54$

74. $\dfrac{14}{-7} + \dfrac{-18}{6 - 15}$

      $= -2 + \dfrac{-18}{-9}$

      $= -2 + 2$

      $= 0$

76. $3(2[12 - 4(8 - 6)])$

      $= 3(2[12 - 4(2)])$

      $= 3(2[12 - 8])$

      $= 3(2[4])$

      $= 24$

78. $8[-13 + 2(5 - 13)]$

$= 8[-13 + 2(-8)]$

$= 8[-13 + (-16)]$

$= 8[-29]$

$= -232$

80. $21 \div [-3 - 5(7 - 9)]$

$= 21 \div [-3 - 5(-2)]$

$= 21 \div [-3 + 10]$

$= 21 \div [7]$

$= 3$

82. $\dfrac{4(-1) + 3(-2) - 23}{-1 - 1 - 13(-1)}$

$= \dfrac{-4 + (-6) - 23}{-1 - 1 + 13}$

$= \dfrac{-33}{11}$

$= -3$

84. $\left[\dfrac{(-7) + 3}{-1 + 5}\right]\left[\dfrac{28 + 4(-10)}{2 - 8}\right]$

$= \left[\dfrac{-4}{4}\right]\left[\dfrac{28 + (-40)}{-6}\right]$

$= [-1]\left[\dfrac{-12}{-6}\right]$

$= [-1][2]$

$= -2$

86. $\dfrac{14 - 2(\frac{8 + 6}{1 - 8}) + 3}{-8 + 5(\frac{2 - 4}{7 - 5}) - 4}$

$= \dfrac{14 - 2(\frac{14}{-7}) + 3}{-8 + 5(\frac{-2}{2}) - 4}$

$= \dfrac{14 - 2(-2) + 3}{-8 + 5(-1) - 4}$

$= \dfrac{14 + 4 + 3}{-8 + (-5) - 4}$

$= \dfrac{21}{-17}$

$= -\dfrac{21}{17}$

88. $\dfrac{-1}{8} \cdot (-\frac{4}{3}) + (\frac{-1}{2}) \div (-\frac{3}{-5})$

$= -\dfrac{1}{\overset{1}{\cancel{8}}}(-\dfrac{4}{3}) + (-\dfrac{1}{\underset{2}{\cancel{2}}})(\dfrac{5}{3})$

$= \dfrac{1}{6} + (-\dfrac{5}{6})$

$= -\dfrac{4}{6}$

$= -\dfrac{2}{3}$

90. $\dfrac{4u + 3v}{w}$

$= \dfrac{4(-4) + 3(8)}{-5}$

$= \dfrac{-16 + 24}{-5}$

$= \dfrac{8}{-5}$

$= -\dfrac{8}{5}$

92. $\dfrac{3u - v + 2w}{6}$

$= \dfrac{3(-4) - 8 + 2(-5)}{6}$

$= \dfrac{-12 - 8 + (-10)}{6}$

$= \dfrac{-30}{6}$

$= -5$

94. $\dfrac{uv + uw - vw}{-v} = \dfrac{(-4)8 + (-4)(-5) - 8(-5)}{-8}$

$= \dfrac{-32 + 20 + 40}{-8}$

$= \dfrac{28}{-8}$

$= -\dfrac{7}{2}$

96.   a)   Since the product of m and n is negative, m and n have unlike signs.

b)   Since the quotient of m and n is positive, m and n have like signs.

98. $18(-4.37) + 29(-2.26) + 46(15.72) + 32(24.85)$

$= -78.66 + (-65.54) + 723.12 + 795.20$

$= \$1374.12$

Problem Set 1.6, pp. 41-42

2.   Identity

6.   Identity

10.   Conditional equation

14.   Conditional equation

18.   Identity

22.   Identity

4.   Identity

8.   Contradiction

12.   Identity

16.   Identity

20.   Contradiction

24. $x + 4 = 9$

       $x + 4 - 4 = 9 - 4$

           $x = 5$

26. $x - 5 = 3$

       $x - 5 + 5 = 3 + 5$

           $x = 8$

28. $2x = 14$

     $\dfrac{2x}{2} = \dfrac{14}{2}$

       $x = 7$

30. $\dfrac{3}{4}x = 5$

     $\dfrac{4}{3} \cdot \dfrac{3}{4}x = \dfrac{4}{3} \cdot 5$

         $x = \dfrac{20}{3}$

32. $y + 6 = 10$

       $y + 6 - 6 = 10 - 6$

           $y = 4$

34. $4t = 12$

     $\dfrac{4t}{4} = \dfrac{12}{4}$

       $t = 3$

36. $r - 3 = -5$

       $r - 3 + 3 = -5 + 3$

           $r = -2$

38. $6s = -18$

     $\dfrac{6s}{6} = \dfrac{-18}{6}$

       $s = -3$

40. $-8w = 40$

     $\dfrac{-8w}{-8} = \dfrac{40}{-8}$

       $w = -5$

42. $5z = 0$

     $\dfrac{5z}{5} = \dfrac{0}{5}$

       $z = 0$

44. $\dfrac{1}{3}x = -6$

     $3 \cdot \dfrac{1}{3}x = 3(-6)$

         $x = -18$

46. $-y = 9$

     $\dfrac{-y}{-1} = \dfrac{9}{-1}$

       $y = -9$

48. $-\dfrac{3}{8}v = -27$

     $-\dfrac{8}{3}\left(-\dfrac{3}{8}v\right) = -\dfrac{8}{3}(-27)$

         $v = 72$

50. $3t - 4 = 11$

       $3t - 4 + 4 = 11 + 4$

           $3t = 15$

         $\dfrac{3t}{3} = \dfrac{15}{3}$

           $t = 5$

52. $2s + 7 = -9$

$2s + 7 - 7 = -9 - 7$

$2s = -16$

$\dfrac{2s}{2} = \dfrac{-16}{2}$

$s = -8$

54. $5x + 1 = 5$

$5x + 1 - 1 = 5 - 1$

$5x = 4$

$\dfrac{5x}{5} = \dfrac{4}{5}$

$x = \dfrac{4}{5}$

56. $4m - 6 = -10$

$4m - 6 + 6 = -10 + 6$

$4m = -4$

$\dfrac{4m}{4} = \dfrac{-4}{4}$

$m = -1$

58. $3n + 19 = 19$

$3n + 19 - 19 = 19 - 19$

$3n = 0$

$\dfrac{3n}{3} = \dfrac{0}{3}$

$n = 0$

60. $-8r - 3 = -3$

$-8r - 3 + 3 = -3 + 3$

$-8r = 0$

$\dfrac{-8r}{-8} = \dfrac{0}{-8}$

$r = 0$

62. $-p - 8 = 12$

$-p - 8 + 8 = 12 + 8$

$-p = 20$

$\dfrac{-p}{-1} = \dfrac{20}{-1}$

$p = -20$

64. $-v + 8 = -6$

$-v + 8 - 8 = -6 - 8$

$-v = -14$

$\dfrac{-v}{-1} = \dfrac{-14}{-1}$

$v = 14$

66. $13 - 8q = 5$

$13 - 8q - 13 = 5 - 13$

$-8q = -8$

$\dfrac{-8q}{-8} = \dfrac{-8}{-8}$

$q = 1$

68. $\qquad -17 - 9y = 16$

$$-17 - 9y + 17 = 16 + 17$$

$$-9y = 33$$

$$\frac{-9y}{-9} = \frac{33}{-9}$$

$$y = -\frac{11}{3}$$

70. $\qquad x + 23 = 38$

$$x + 23 - 23 = 38 - 23$$

$$x = 15$$

72. $\quad \frac{3}{5}y = 8$

$$\frac{5}{3} \cdot \frac{3}{5}y = \frac{5}{3} \cdot 8$$

$$y = \frac{40}{3}$$

74. $\qquad 4t + 19 = 51$

$$4t + 19 - 19 = 51 - 19$$

$$4t = 32$$

$$\frac{4t}{4} = \frac{32}{4}$$

$$t = 8$$

76. $0.16r = 13.6$

$$\frac{0.16r}{0.16} = \frac{13.6}{0.16}$$

$$r = 85$$

78. $\qquad 175 + 0.15s = 415$

$$175 + 0.15s - 175 = 415 - 175$$

$$0.15s = 240$$

$$\frac{0.15s}{0.15} = \frac{240}{0.15}$$

$$s = \$1600$$

80. a) $3s = 117.9748$

$$\frac{3s}{3} = \frac{117.9748}{3}$$

$$s \approx 39.3249 \text{ cm}$$

b) $4s = 117.9748$

$$\frac{4s}{4} = \frac{117.9748}{4}$$

$$s = 29.4937 \text{ cm}$$

c) $5s = 117.9748$

$$\frac{5s}{5} = \frac{117.9748}{5}$$

$$s \approx 23.5950 \text{ cm}$$

d) $\pi s = 117.9748$

$$\frac{\pi s}{\pi} = \frac{117.9748}{\pi}$$

$$s \approx 37.5525 \text{ cm}$$

NOTES

CHAPTER 2

EXPONENTS AND POLYNOMIALS

Problem Set 2.1, pp. 51-52

2. Base = 9, exponent = 6   4. Base = -5, exponent = 4

6. Base = 5, exponent = 4   8. Base = 3y, exponent = 2

10. Base = y, exponent = 2   12. Base = -t, exponent = 2

14. Base = t, exponent = 2   16. Base = r + 3, exponent = 8

18. Base = ab, exponent = n − 1   20. Base = $\frac{a}{b}$, exponent = 3n

22. $5^2 = 5 \cdot 5 = 25$ 24. $(-7)^2 = (-7)(-7) = 49$

$2^5 = 2 \cdot 2 \cdot 2 \cdot 2 \cdot 2 = 32$ $-7^2 = -(7 \cdot 7) = -49$

26. $(-9)^3 = (-9)(-9)(-9) = -729$ 28. $(-2)^4 = (-2)(-2)(-2)(-2) = 16$

$-9^3 = -(9 \cdot 9 \cdot 9) = -729$ $-2^4 = -(2 \cdot 2 \cdot 2 \cdot 2) = -16$

30. $(2 \cdot 5)^3 = (10)^3 = 1000$ 32. $9(-\frac{2}{3})^4 = 9(\frac{16}{81}) = \frac{16}{9}$

$2 \cdot 5^3 = 2 \cdot 125 = 250$

$-9(\frac{2}{3})^4 = -9(\frac{16}{81}) = -\frac{16}{9}$

34. $(3 + 5)^2 = (8)^2 = 64$ 36. $(4 - 2)^2 = (2)^2 = 4$

$3^2 + 5^2 = 9 + 25 = 34$ $4^2 - 2^2 = 16 - 4 = 12$

38. $(3^2)^3 = (9)^3 = 729$ 40. $(x + 4)(x + 4)(x + 4)$

$3^{(2^3)} = 3^8 = 6561$ $= (x + 4)^3$

42. $a \cdot a \cdot a \cdot b - 17 \cdot a \cdot a \cdot b \cdot b \cdot b \cdot b = a^3 b - 17a^2 b^4$

44. $(6s)(6s)(6s)(6s)(6s) = (6s)^5$

46. $\frac{n}{9} \cdot \frac{n}{9} \cdot \frac{n}{9} \cdot \frac{n}{9} = (\frac{n}{9})^4$

48. $V = P(1 + r)^t$

        $V = 50,000(1 + 0.20)^t$

        $V = 50,000(1.2)^t$

a) $V = 50,000(1.2)^1 = 50,000(1.2) = \$60,000$

b) $V = 50,000(1.2)^2 = 50,000(1.44) = \$72,000$

c) $V = 50,000(1.2)^3 = 50,000(1.728) = \$86,400$

d) $V = 50,000(1.2)^4 = 50,000(2.0736) = \$103,680$

50. a)

	Game 1
Choice 1	Home
Choice 2	Visiting

b)

	Game 1	Game 2
Choice 1	Home	Home
Choice 2	Home	Visiting
Choice 3	Visiting	Home
Choice 4	Visiting	Visiting

c)

	Game 1	Game 2	Game 3
Choice 1	Home	Home	Home
Choice 2	Home	Home	Visiting
Choice 3	Home	Visiting	Home
Choice 4	Visiting	Home	Home
Choice 5	Home	Visiting	Visiting
Choice 6	Visiting	Home	Visiting
Choice 7	Visiting	Visiting	Home
Choice 8	Visiting	Visiting	Visiting

$2^{10} = 1024$ ways, 2^n ways

52. $V = P(1 + r)^t$

$V = 3500(1 + 0.1225)^{18} = 3500(1.1225)^{18} = \$28,017.05$

Problem Set 2.2, pp. 57-58

2. $x^4 \cdot x^6 = x^{4+6} = x^{10}$

4. $4^2 \cdot 4^3 = 4^{2+3} = 4^5$

6. $6^5 \cdot 6^3 \cdot 6 = 6^{5+3+1} = 6^9$

8. $7r^4 \cdot r^9 = 7r^{4+9} = 7r^{13}$

10. $t^n t^3 = t^{n+3}$

12. $(y - 4)^7 (y - 4)^3 = (y - 4)^{7+3}$

$= (y - 4)^{10}$

14. $\dfrac{y^8}{y^2} = y^{8-2} = y^6$

16. $\dfrac{8^6}{8^2} = 8^{6-2} = 8^4$

18. $\dfrac{5t^9}{t^4} = 5t^{9-4} = 5t^5$

20. $\dfrac{a^5 b^7}{a^3 b} = a^{5-3} b^{7-1} = a^2 b^6$

22. $\dfrac{r^{n+1}}{r} = \dfrac{r^{n+1}}{r^1} = r^{n+1-1} = r^n$

24. $\dfrac{(6m + 1)^{201}}{(6m + 1)^{200}} = (6m + 1)^{201-200} = (6m + 1)^1 = 6m + 1$

26. $(x^3)^5 = x^{3 \cdot 5} = x^{15}$

28. $(3^2)^9 = 3^{2 \cdot 9} = 3^{18}$

30. $(b^{15})^{2n} = b^{15 \cdot 2n} = b^{30n}$

32. $(r^{q-1})^3 = r^{(q-1)3} = r^{3q-3}$

34. $(4p)^3 = 4^3 p^3 = 64p^3$

36. $(-8mn)^2 = (-8)^2 m^2 n^2$

$= 64m^2 n^2$

38. $(5a)^4 = 5^4 a^4 = 625a^4$

40. $(10t)^n = 10^n t^n$

42. $(\frac{2}{5})^2 = \dfrac{2^2}{5^2} = \dfrac{4}{25}$

44. $(\frac{-3}{z})^3 = \dfrac{(-3)^3}{z^3} = \dfrac{-27}{z^3}$

46. $(3x^2)(8x) = (3 \cdot 8)(x^2 \cdot x) = 24x^3$

48. $(\frac{2x^3}{y^4})^2 = \dfrac{(2x^3)^2}{(y^4)^2} = \dfrac{2^2 (x^3)^2}{y^8} = \dfrac{4x^6}{y^8}$

50. $(4p^5)^3 = 4^3 (p^5)^3 = 64p^{15}$

52. $(4st^4)(-7s^2 t^6) = 4(-7)(ss^2)(t^4 t^6) = -28s^3 t^{10}$

54. $(-2p^6 q)^5 = (-2)^5 (p^6)^5 q^5 = -32p^{30} q^5$

56. $\dfrac{-16(a^2b^{10})^2}{8ab^5} = \dfrac{-16(a^2)^2(b^{10})^2}{8ab^5} = \dfrac{-16a^4b^{20}}{8ab^5} = -2a^3b^{15}$

58. $\dfrac{3b^4}{5m}(\dfrac{3m^3}{b^6})^3 = \dfrac{3b^4}{5m} \cdot \dfrac{3^3(m^3)^3}{(b^6)^3} = \dfrac{3b^4}{5m} \cdot \dfrac{27m^9}{b^{18}} = \dfrac{81m^8}{5b^{14}}$

60. $-x^2(-5x^3)^2(2x^4)^3 = -x^2(-5)^2(x^3)^2 2^3(x^4)^3$

$\qquad\qquad\qquad\qquad = -x^2(25)x^6(8)x^{12}$

$\qquad\qquad\qquad\qquad = -200x^{20}$

62. $(\dfrac{3x^5}{4y^3})^4(\dfrac{4x^2}{y^4z^0})^3 = \dfrac{3^4(x^5)^4}{4^4(y^3)^4} \cdot \dfrac{4^3(x^2)^3}{(y^4)^3 \cdot 1^3}$

$\qquad\qquad\qquad = \dfrac{81x^{20}}{4^4y^{12}} \cdot \dfrac{4^3x^6}{y^{12}}$

$\qquad\qquad\qquad = \dfrac{81x^{26}}{4y^{24}}$

64. $(-8ab^2c^3)(-3a^2bc^5)(2abc^7) = (-8)(-3)2(aa^2a)(b^2bb)(c^3c^5c^7)$

$\qquad\qquad\qquad\qquad\qquad\qquad = 48a^4b^4c^{15}$

66. $(9r^2s)(-3r^4s^4)^3(r^8s^{10})^4 = 9r^2s(-3)^3(r^4)^3(s^4)^3(r^8)^4(s^{10})^4$

$\qquad\qquad\qquad\qquad\qquad = 9r^2s(-27)r^{12}s^{12}r^{32}s^{40}$

$\qquad\qquad\qquad\qquad\qquad = -243r^{46}s^{53}$

68. $\dfrac{-m^2}{6n^6}(\dfrac{-2n^4}{m^2})^6 = \dfrac{-m^2}{6n^6} \cdot \dfrac{(-2)^6(n^4)^6}{(m^2)^6}$

$\qquad\qquad\qquad = \dfrac{-m^2}{6n^6} \cdot \dfrac{64n^{24}}{m^{12}}$

$\qquad\qquad\qquad = -\dfrac{32n^{18}}{3m^{10}}$

70. $\dfrac{(3w^4)^3(-5w^3)^2}{6w(10w^2)^3} = \dfrac{3^3(w^4)^3(-5)^2(w^3)^2}{6w(10^3)(w^2)^3}$

$\qquad\qquad\qquad = \dfrac{27w^{12}(25)w^6}{6w(1000)w^6}$

$\qquad\qquad\qquad = \dfrac{9w^{11}}{80}$

72. a) $99^0 = 1$ b) $(\frac{2}{5})^0 = 1$

     c) $4x^0 = 4 \cdot 1 = 4$ d) $(-7)^0 = 1$

     e) $-7^0 = -(7^0) = -1$

74. $81^{60} = (3^4)^{60} = 3^{4 \cdot 60} = 3^{240}$

76. a) After 20 min, the culture contains $7 \cdot 2 = 7 \cdot 2^1$ cells.

       After 40 min, the culture contains $(7 \cdot 2^1) \cdot 2 = 7 \cdot 2^2$ cells.

       After 60 min, the culture contains $(7 \cdot 2^2) \cdot 2 = 7 \cdot 2^3$ cells.

     b) After 6 hr (18 20-min periods), the culture contains $7 \cdot 2^{18}$ cells.

78. $$\left(\frac{a}{b}\right)^n = \underbrace{\left(\frac{a}{b}\right)\left(\frac{a}{b}\right) \ldots \ldots \left(\frac{a}{b}\right)}_{n \text{ factors of } \frac{a}{b}} = \frac{\overbrace{a \cdot a \ldots \ldots a}^{n \text{ factors of } a}}{\underbrace{b \cdot b \ldots \ldots b}_{n \text{ factors of } b}} = \frac{a^n}{b^n}$$

80. a) $\left(\dfrac{-3a^{555}}{b^{747}}\right)^{15} = \dfrac{(-3)^{15}(a^{555})^{15}}{(b^{747})^{15}} = -\dfrac{3^{15}a^{8325}}{b^{11,205}}$

     b) $28r^{3817}s^{1729}(-2r^{987}s^{613})^{18} = 28r^{3817}s^{1729}(-2)^{18}(r^{987})^{18}(s^{613})^{18}$

                     $= 7 \cdot 2^2 r^{3817}s^{1729}2^{18}r^{17,766}s^{11,034}$

                     $= 7 \cdot 2^{20}r^{21,583}s^{12,763}$

Problem Set 2.3, pp. 61-62

2. Coefficient = 3, degree = 7

4. Coefficient = 1 (since $y^3 = 1y^3$), degree = 3

6. Coefficient = 7, degree = 1 (since $7z = 7z^1$)

8. Coefficient = -1 (since $-t^6 = -1t^6$), degree = 6

10. Coefficient = 16, degree = 0 (since $16 = 16x^0$)

12. Coefficient = $-\frac{4}{5}$, degree = 1 (since $-\frac{4}{5}r = -\frac{4}{5}r^1$)

14. $x^2 + 8x + 6$, degree = 2, trinomial

16. $x^2 - 4x + 2$, degree = 2, trinomial

18. $x + 3$, degree = 1, binomial

20. $2y + 5$, degree = 1, binomial

22. $r^2 - 4$, degree = 2, binomial

24. $-p^2 + 3$, degree = 2, binomial

26. $99t^{100}$, degree = 100, monomial

28. $8m^4 + m^2 - 1$, degree = 4, trinomial

30. $q^5 - 2q^3 - q^2 + q$, degree = 5

32. $3x$, degree = 1, monomial

34. 3, degree = 0, monomial

36. $x + 4$

       $= 3 + 4$

       $= 7$

38. $5x + 3$

       $= 5(-1) + 3$

       $= -5 + 3$

       $= -2$

40. $y^2 + 6$

       $= 2^2 + 6$

       $= 4 + 6$

       $= 10$

42. $4y^2 - 9y + 12$

       $= 4(4)^2 - 9(4) + 12$

       $= 4(16) - 36 + 12$

       $= 48 - 36 + 12$

       $= 24$

44. $z^3 - 2z^2 + 5z - 8$

       $= 0^3 - 2(0)^2 + 5(0) - 8$

       $= 0 - 0 + 0 - 8$

       $= -8$

46. $z^5 - z^3$

       $= (-3)^5 - (-3)^3$

       $= -243 - (-27)$

       $= -216$

48. $2r^3 - 5r^2 + 4r + 2$

      $= 2(\frac{1}{2})^3 - 5(\frac{1}{2})^2 + 4(\frac{1}{2}) + 2$

      $= 2(\frac{1}{8}) - 5(\frac{1}{4}) + 2 + 2$

      $= \frac{1}{4} - \frac{5}{4} + \frac{8}{4} + \frac{8}{4}$

      $= \frac{12}{4}$

      $= 3$

50. $-t^2 + 7t + 13$

      $= -5^2 + 7(5) + 13$

      $= -25 + 35 + 13$

      $= 23$

52. $-t^4 + t^2 - t + 3$

      $= -(-2)^4 + (-2)^2 - (-2) + 3$

      $= -16 + 4 + 2 + 3$

      $= -7$

54. $h = -16t^2 + rt + s$

      $h = -16t^2 + 80t + 96$

    a) $h = -16(1)^2 + 80(1) + 96 = -16 + 80 + 96 = 160$ ft

    b) $h = -16(2)^2 + 80(2) + 96 = -64 + 160 + 96 = 192$ ft

    c) $h = -16(3)^2 + 80(3) + 96 = -144 + 240 + 96 = 192$ ft

    d) $h = -16(4)^2 + 80(4) + 96 = -256 + 320 + 96 = 160$ ft

    e) $h = -16(5)^2 + 80(5) + 96 = -400 + 400 + 96 = 96$ ft

    f) $h = -16(6)^2 + 80(6) + 96 = -576 + 480 + 96 = 0$ ft

$t = 2, h = 192$ $t = 3, h = 192$

$t = 1, h = 160$ $t = 4, h = 160$

$t = 5, h = 96$

96 ft

$t = 6, h = 0$

Maximum height occurs at $t = 2.5$ sec, and is

$h = -16(2.5)^2 + 80(2.5) + 96 = -100 + 200 + 96 = 196$ ft

56. a) Since s = 0, the formula is

$$h = -16t^2 + 80t + 0$$

$$h = -16t^2 + 80t.$$

 b) Since r = 0, the formula is

$$h = -16t^2 + 0t + 96$$

$$h = -16t^2 + 96.$$

58. $d = 0.044r^2 + 1.1r$

$d = 0.044(55)^2 + 1.1(55)$

$= 0.044(3025) + 60.5$

$= 133.1 + 60.5$

$= 193.6$ ft

Problem Set 2.4, pp. 66-67

2. Like terms

4. Like terms

6. Unlike terms

8. Unlike terms

10. Like terms

12. Unlike terms

14. Unlike terms

16. Unlike terms

18. $3x + 2x = (3 + 2)x$

$= 5x$

20. $8y^2 + 6y^2 + y^2 = (8 + 6 + 1)y^2$

$= 15y^2$

22. $-4z + 7z = (-4 + 7)z$

$= 3z$

24. $rs - 7rs = (1 - 7)rs$

$= -6rs$

26. $-11p + 11p + 6q$

$= (-11 + 11)p + 6q$

$= 0p + 6q$

$= 6q$

28. $-3x^2y^2 - 4x^2y^2$

$= (-3 - 4)x^2y^2$

$= -7x^2y^2$

30. $\quad 5m^3 + m^2 + 3m^3$

$= 5m^3 + 3m^3 + m^2$

$= (5 + 3)m^3 + m^2$

$= 8m^3 + m^2$

32. $\quad -5cd^2 - (-cd^2) + 4c^2d$

$= -5cd^2 + cd^2 + 4c^2d$

$= (-5 + 1)cd^2 + 4c^2d$

$= -4cd^2 + 4c^2d$

34. $\quad 4(x - 5) = 4 \cdot x - 4 \cdot 5 = 4x - 20$

36. $\quad 7(3x - 2) = 7 \cdot 3x - 7 \cdot 2 = 21x - 14$

38. $\quad -3(4y - 1) = (-3)4y + (-3)(-1) = -12y + 3$

40. $\quad -(r + 6) = (-1)r + (-1)6 = -r + (-6) = -r - 6$

42. $\quad -(-t + 3) = (-1)(-t) + (-1)3 = t + (-3) = t - 3$

44. $\quad -2(-s^2 - 5s - 2) = (-2)(-s^2) + (-2)(-5s) + (-2)(-2)$

$\qquad\qquad\qquad = 2s^2 + 10s + 4$

46. $\quad -(y^2 - 3y - 9) = (-1)y^2 + (-1)(-3y) + (-1)(-9)$

$\qquad\qquad\qquad = -y^2 + 3y + 9$

48. $\quad (8x^2 - 3x + 2) + (2x^2 + 11x + 1)$

$= 8x^2 - 3x + 2 + 2x^2 + 11x + 1$

$= (8x^2 + 2x^2) + (-3x + 11x) + (2 + 1)$

$= 10x^2 + 8x + 3$

$$\begin{array}{r} 8x^2 - 3x + 2 \\ \underline{2x^2 + 11x + 1} \\ 10x^2 + 8x + 3 \end{array}$$

50. $\quad (8x^3 + 3x^2 - 2) + (5x^3 - 2x^2 - x + 4)$

$= (8x^3 + 5x^3) + (3x^2 - 2x^2) + (-x) + (-2 + 4)$

$= 13x^3 + x^2 - x + 2$

$$\begin{array}{r} 8x^3 + 3x^2 \qquad - 2 \\ \underline{5x^3 - 2x^2 - x + 4} \\ 13x^3 + x^2 - x + 2 \end{array}$$

52. $\quad (9m^2 - 6mn + n^2) + (5m^2 + 3mn + 7n^2)$

$= (9m^2 + 5m^2) + (-6mn + 3mn) + (n^2 + 7n^2)$

$= 14m^2 + (-3mn) + 8n^2$

$= 14m^2 - 3mn + 8n^2$

$$\begin{array}{r} 9m^2 - 6mn + n^2 \\ \underline{5m^2 + 3mn + 7n^2} \\ 14m^2 - 3mn + 8n^2 \end{array}$$

54. $(10x^2 + 8x + 2) - (4x^2 + 6x - 3)$

$= 10x^2 + 8x + 2 - 4x^2 - 6x + 3$

$= 10x^2 - 4x^2 + 8x - 6x + 2 + 3$

$= 6x^2 + 2x + 5$

$$\begin{array}{r} 10x^2 + 8x + 2 \\ \underline{4x^2 + 6x - 3} \\ 6x^2 + 2x + 5 \end{array}$$

56. $(8x^3 + 3x^2 - 2) - (5x^3 - 2x^2 - x + 4)$

$= 8x^3 + 3x^2 - 2 - 5x^3 + 2x^2 + x - 4$

$= 8x^3 - 5x^3 + 3x^2 + 2x^2 + x - 2 - 4$

$= 3x^3 + 5x^2 + x - 6$

$$\begin{array}{r} 8x^3 + 3x^2 \qquad - 2 \\ \underline{5x^3 - 2x^2 - x + 4} \\ 3x^3 + 5x^2 + x - 6 \end{array}$$

58. $(2r^3 + r^2s^2 - 7s^3) - (r^3 + r^2s^2 + 7s^3)$

$= 2r^3 + r^2s^2 - 7s^3 - r^3 - r^2s^2 - 7s^3$

$= 2r^3 - r^3 + r^2s^2 - r^2s^2 - 7s^3 - 7s^3$

$= r^3 - 14s^3$

$$\begin{array}{r} 2r^3 + r^2s^2 - 7s^3 \\ \underline{r^3 + r^2s^2 + 7s^3} \\ r^3 \qquad - 14s^3 \end{array}$$

60. $5(2p^3 + 3p^2 - p) - 4(5p^2 - 4p + 7)$

$= 10p^3 + 15p^2 - 5p - 20p^2 + 16p - 28$

$= 10p^3 - 5p^2 + 11p - 28$

62. $-7(a - 2b + c) + 5(2a - b + c) - (a + b - 2c)$

$= -7a + 14b - 7c + 10a - 5b + 5c - a - b + 2c$

$= -7a + 10a - a + 14b - 5b - b - 7c + 5c + 2c$

$= 2a + 8b$

64. $2 + 9[(r + 5) - (r - 7)]$

$= 2 + 9[r + 5 - r + 7]$

$= 2 + 9[12]$

$= 2 + 108$

$= 110$

66. $7u - [2u - 3(5u + 2)]$

$= 7u - [2u - 15u - 6]$

$= 7u - [-13u - 6]$

$= 7u + 13u + 6$

$= 20u + 6$

68. $3[t - (2t + t - 6) + 5]$

= $3[t - (3t - 6) + 5]$

= $3[t - 3t + 6 + 5]$

= $3[-2t + 11]$

= $-6t + 33$

70. $n - (n - [m - (3m + n)] - 4m)$

= $n - (n - [m - 3m - n] - 4m)$

= $n - (n - [-2m - n] - 4m)$

= $n - (n + 2m + n - 4m)$

= $n - (-2m + 2n)$

= $n + 2m - 2n$

= $2m - n$

72. $(7x^3y)(2xy^2) - (x^2y^2)(5x^2y)$

= $14x^4y^3 - 5x^4y^3$

= $9x^4y^3$

74. $14 - 4(a - 8a) + 3 - (12 + 7a) - (a)(-5)$

= $14 - 4a + 32a + 3 - 12 - 7a + 5a$

= $26a + 5$

76. $(9)(4a^2) - (11a^2)(-5) + (-6a^2)(-3) + (15b)(3)$

= $36a^2 + 55a^2 + 18a^2 + 45b$

= $109a^2 + 45b$

78. $(-1)(-17a^3) - (-a^3)(-8) + (19a^3)(-2) - (-6)(-3a^3)$

= $17a^3 - 8a^3 - 38a^3 - 18a^3$

= $-47a^3$

80. $n + (n + 2) + (n + 4) = 3n + 6$

82. $10m + 5(m - 4) + 25[3(m - 4)]$

= $10m + 5m - 20 + 25[3m - 12]$

= $10m + 5m - 20 + 75m - 300$

= $90m - 320$

84. $p + 20\%$ of $p = 1p + 0.2p = 1.2p$ dollars

86. $(-0.67)(4.7a) - (-3.5a)(16) + (7.8a)(-5.6) - (-8.2a)(-2.9)$

   $= -3.149a + 56a - 43.68a - 23.78a$

   $= -14.609a$

Problem Set 2.5, pp. 73-74

2. $3 \cdot 8x = 24x$

4. $5x^3 \cdot 2 = 2 \cdot 5x^3 = 10x^3$

6. $9y \cdot y = 9y^2$

8. $(6y^2)(7y^5) = (6 \cdot 7)(y^2 y^5)$

   $= 42y^7$

10. $(-3r^3)(-4r^2 s^6) = (-3)(-4)(r^3 r^2)s^6 = 12r^5 s^6$

12. $(2r^2 s^5)(-r^3 s^3) = 2(-1)(r^2 r^3)(s^5 s^3) = -2r^5 s^8$

14. $(8x)(9y) = (8 \cdot 9)(x \cdot y) = 72xy$

16. $x(x - 4) = x \cdot x - x \cdot 4 = x^2 - 4x$

18. $3x^2(4x + 5) = 3x^2 \cdot 4x + 3x^2 \cdot 5 = 12x^3 + 15x^2$

20. $6r(r^3 + 2r^2 - 1) = 6r \cdot r^3 + 6r \cdot 2r^2 - 6r \cdot 1 = 6r^4 + 12r^3 - 6r$

22. $2r^2(r^2 - 5rs + 8s^2) = 2r^2 \cdot r^2 - 2r^2 \cdot 5rs + 2r^2 \cdot 8s^2$

   $= 2r^4 - 10r^3 s + 16r^2 s^2$

24. $-m(m^2 + mn - n^2) = (-m)m^2 + (-m)mn - (-m)n^2$

   $= -m^3 - m^2 n + mn^2$

26. $st^3(s^3 - s^2 t^3 - t^4) = st^3 \cdot s^3 - st^3 \cdot s^2 t^3 - st^3 \cdot t^4$

   $= s^4 t^3 - s^3 t^6 - st^7$

28. $(x + 3)(x + 5) = x \cdot x + x \cdot 5 + 3 \cdot x + 3 \cdot 5$

   $= x^2 + 5x + 3x + 15$

   $= x^2 + 8x + 15$

30. $(4x - 3)(x^2 + 5x - 7) = 4x \cdot x^2 + 4x \cdot 5x - 4x \cdot 7 - 3 \cdot x^2 - 3 \cdot 5x - 3$

   $= 4x^3 + 20x^2 - 28x - 3x^2 - 15x + 21$

   $= 4x^3 + 17x^2 - 43x + 21$

32. $(8r - 3)(5r^3 + 2r^2 - 7r - 4)$

$= 8r \cdot 5r^3 + 8r \cdot 2r^2 - 8r \cdot 7r - 8r \cdot 4 - 3 \cdot 5r^3 - 3 \cdot 2r^2 - 3(-7r) - 3(-4)$

$= 40r^4 + 16r^3 - 56r^2 - 32r - 15r^3 - 6r^2 + 21r + 12$

$= 40r^4 + r^3 - 62r^2 - 11r + 12$

34. $(b - 5)(b + 3)(b - 1) = (b^2 + 3b - 5b - 15)(b - 1)$

$= (b^2 - 2b - 15)(b - 1)$

$= b^3 - b^2 - 2b^2 + 2b - 15b + 15$

$= b^3 - 3b^2 - 13b + 15$

36. $(3p - 2)^3 = (3p - 2)(3p - 2)(3p - 2)$

$= (9p^2 - 6p - 6p + 4)(3p - 2)$

$= (9p^2 - 12p + 4)(3p - 2)$

$= 27p^3 - 18p^2 - 36p^2 + 24p + 12p - 8$

$= 27p^3 - 54p^2 + 36p - 8$

38. $y - 8$

$\underline{y + 1}$

$y^2 - 8y$

$\underline{\qquad y - 8}$

$y^2 - 7y - 8$

40. $x^2 + 6x - 2$

$\underline{\qquad 3x - 4}$

$3x^3 + 18x^2 - 6x$

$\underline{\qquad - 4x^2 - 24x + 8}$

$3x^3 + 14x^2 - 30x + 8$

42. $4m^3 + 5m - 2$

$\underline{\qquad 2m - 3}$

$8m^4 \qquad + 10m^2 - 4m$

$\underline{\quad - 12m^3 \qquad -15m + 6}$

$8m^4 - 12m^3 + 10m^2 -19m + 6$

44. $3a^2 - 5ab + b^2$

$\underline{\qquad 3a - 4b}$

$9a^3 - 15a^2b + 3ab^2$

$\underline{\quad - 12a^2b + 20ab^2 - 4b^3}$

$9a^3 - 27a^2b + 23ab^2 - 4b^3$

46. $6m - n + 3$

$\underline{2m + 4n - 5}$

$12m^2 - 2mn + 6m$

$24mn \qquad - 4n^2 + 12n$

$\underline{\qquad\qquad - 30m \qquad + 5n - 15}$

$12m^2 + 22mn - 24m - 4n^2 + 17n - 15$

48. $(x + 4)(x + 7) = x^2 + 11x + 28$

50. $(x + 8)(x - 3) = x^2 + 5x - 24$

52. $(5r - 4)(r + 2) = 5r^2 + 6r - 8$

54. $(2s - 3)(6s - 11) = 12s^2 - 40s + 33$

56. $(5p + 6q)(4p - 3q) = 20p^2 + 9pq - 18q^2$

58. $(t^2 - 1)(t^2 - 16) = t^4 - 17t^2 + 16$

60. $(4a^2 - 5b)(3a^2 + b) = 12a^4 - 11a^2b - 5b^2$

62. $(y^3 - 9z)(y^3 + z) = y^6 - 8y^3z - 9z^2$

64. $(x^n + 13)(x^n + 15) = x^{2n} + 28x^n + 195$

66. $(x + 3)^2 = \qquad x^2 \qquad + \qquad 2 \cdot x \cdot 3 \qquad + \qquad 3^2 \qquad = x^2 + 6x + 9$

square	twice	square
first	product	last
term	of terms	term

68. $(2y - 5)^2 = (2y)^2 + 2(2y)(-5) + (-5)^2 = 4y^2 - 20y + 25$

70. $(p + 4)(p - 4) = p^2 - 4^2 = p^2 - 16$

$(a + b)(a - b) = a^2 - b^2$

72. $(7r + 2s)(7r - 2s) = (7r)^2 - (2s)^2 = 49r^2 - 4s^2$

74. $(5m - 3n)^2 = (5m)^2 + 2(5m)(-3n) + (-3n)^2 = 25m^2 - 30mn + 9n^2$

76. $(t^{3k} - 6)(t^{3k} + 6) = (t^{3k})^2 - 6^2 = t^{6k} - 36$

78. $m^2[m - (m + 8)(m - 1)]$

$= m^2[m - (m^2 + 7m - 8)]$

$= m^2[m - m^2 - 7m + 8]$

$= m^2[-m^2 - 6m + 8]$

$= -m^4 - 6m^3 + 8m^2$

80. $-t(t - 4[2t - 3(t + 2)] + 8)$

$= -t(t - 4[2t - 3t - 6] + 8)$

$= -t(t - 4[-t - 6] + 8)$

$= -t(t + 4t + 24 + 8)$

$= -t(5t + 32)$

$= -5t^2 - 32t$

82. $6pq^2 - 2[q - q(pq - p)] - pq$

$= 6pq^2 - 2[q - pq^2 + pq] - pq$

$= 6pq^2 - 2q + 2pq^2 - 2pq - pq$

$= 8pq^2 - 3pq - 2q$

84. $n(n + 1)(n + 2) = n(n^2 + 3n + 2) = n^3 + 3n^2 + 2n$

86. $\dfrac{n(n - 1)}{2} = \dfrac{15(15 - 1)}{2} = \dfrac{15(14)}{2} = 15(7) = 105$ matches

88. $3 \cdot \dfrac{n(n - 1)}{2} = \dfrac{3n(n - 1)}{2}$ matches

90. $(10^6 + 2)(10^6 - 2) = (10^6)^2 - 2^2 = 10^{12} - 4 \approx 10^{12}$

Problem Set 2.6, pp. 79-80

2. $35 = 5 \cdot 7$

4. $18 = 2 \cdot 3^2$

6. $60 = 2^2 \cdot 3 \cdot 5$

8. $540 = 2^2 \cdot 3^3 \cdot 5$

10. GCF = 3

12. GCF = 4

14. $18 = 2 \cdot 3^2$

    $12 = 2^2 \cdot 3$

    GCF $= 2 \cdot 3 = 6$

16. $24 = 2^3 \cdot 3$

    $36 = 2^2 \cdot 3^2$

    $48 = 2^4 \cdot 3$

    GCF $= 2^2 \cdot 3 = 12$

18. GCF $= 4x^3$

20. GCF $= 11$

22. GCF $= 4$

24. GCF $= 1$

26. GCF $= 5m^4n^2$

28. $4x + 20 = 4 \cdot x + 4 \cdot 5$

    $= 4(x + 5)$

30. $11a + 11b = 11(a + b)$

32. $35m - 45n = 5 \cdot 7m - 5 \cdot 9n$

    $= 5(7m - 9n)$

34. $6y - 25$ is prime

36. $8x^2 + 8 = 8 \cdot x^2 + 8 \cdot 1$

    $= 8(x^2 + 1)$

38. $9x^3 + 12x^2 = 3x^2 \cdot 3x + 3x^2 \cdot 4$

    $= 3x^2(3x + 4)$

40. $a^6 + a = a \cdot a^5 + a \cdot 1$

    $= a(a^5 + 1)$

42. $24a^2 - 18ab = 6a \cdot 4a - 6a \cdot 3b = 6a(4a - 3b)$

44. $-3p^7 + 12p^5 - 9p^3 = 3p^3(-p^4) + 3p^3 \cdot 4p^2 + 3p^3(-3)$

    $= 3p^3(-p^4 + 4p^2 - 3)$

46. $24r^6s^3 - 36r^4s^4 - 48rs = 6rs \cdot 4r^5s^2 - 6rs \cdot 6r^3s^3 - 6rs \cdot 8$

    $= 6rs(4r^5s^2 - 6r^3s^3 - 8)$

48. $100z^{75} - 10z^{50} + 40z^{25} = 10z^{25} \cdot 10z^{50} - 10z^{25} \cdot z^{25} + 10z^{25} \cdot 4$

    $= 10z^{25}(10z^{50} - z^{25} + 4)$

50. $-15r^2s^2t^2 - 25r^4s^2t^3 = (-5r^2s^2t^2)3 + (-5r^2s^2t^2)5r^2t$

    $= -5r^2s^2t^2(3 + 5r^2t)$

52. $\pi x^2y^2z - \pi xy^2z - \pi xy^2z^2 = \pi xyz \cdot xy - \pi xyz \cdot y - \pi xyz \cdot yz$

    $= \pi xyz(xy - y - yz)$

54. $7x(4m + n) - y(4m + n) = (4m + n)(7x - y)$

56. $x^2(x^2 + 4) + 4(x^2 + 4) = (x^2 + 4)(x^2 + 4) = (x^2 + 4)^2$

58. $z^{n+4} + z^{n+2} = z^{n+2} \cdot z^2 + z^{n+2} \cdot 1 = z^{n+2}(z^2 + 1)$

60. $x^2 - 16 = x^2 - 4^2 = (x - 4)(x + 4)$

62. $9x^2 - 16 = (3x)^2 - 4^2 = (3x - 4)(3x + 4)$

64. $25m^2 - 49n^2 = (5m)^2 - (7n)^2 = (5m - 7n)(5m + 7n)$

66. $x^2 + 16$ is prime

68. $100r^2s^2 - 81 = (10rs)^2 - 9^2 = (10rs - 9)(10rs + 9)$

70. $(v - 2)^2 - 4 = (v - 2)^2 - 2^2 = [(v - 2) - 2][(v - 2) + 2]$

$$= (v - 4)v$$

72. $a^2 - (b + c)^2 = [a - (b + c)][a + (b + c)] = (a - b - c)(a + b + c)$

74. $(4x - 3)^2 - (2x + 1)^2 = [(4x - 3) - (2x + 1)][(4x - 3) + (2x + 1)]$

$$= (4x - 3 - 2x - 1)(4x - 3 + 2x + 1)$$

$$= (2x - 4)(6x - 2)$$

76. $6x^2 - 54 = 6(x^2 - 9) = 6(x - 3)(x + 3)$

78. $r^3 - 49r = r(r^2 - 49) = r(r - 7)(r + 7)$

80. $t^4 - 81 = (t^2)^2 - 9^2 = (t^2 - 9)(t^2 + 9) = (t - 3)(t + 3)(t^2 + 9)$

82. $15x^4 - 15y^4 = 15(x^4 - y^4) = 15(x^2 - y^2)(x^2 + y^2)$

$$= 15(x - y)(x + y)(x^2 + y^2)$$

84. $x^4y^4 - 625 = (x^2y^2)^2 - 25^2$

$$= (x^2y^2 - 25)(x^2y^2 + 25)$$

$$= (xy - 5)(xy + 5)(x^2y^2 + 25)$$

86. $66rs - 66r^3s^3 = 66rs(1 - r^2s^2)$

$$= 66rs(1 - rs)(1 + rs)$$

88. $m^6 - 9 = (m^3)^2 - 3^2 = (m^3 - 3)(m^3 + 3)$

90. $1 - 16x^4 = 1^2 - (4x^2)^2 = (1 - 4x^2)(1 + 4x^2)$

$$= (1 - 2x)(1 + 2x)(1 + 4x^2)$$

92. $4p^6q - 64p^2q = 4p^2q(p^4 - 16)$

$$= 4p^2q(p^2 - 4)(p^2 + 4)$$

$$= 4p^2q(p - 2)(p + 2)(p^2 + 4)$$

94. $75^2 - 25^2 = (75 - 25)(75 + 25) = (50)(100) = 5000$

96. $\pi R^2 h - \pi r^2 h = \pi h(R^2 - r^2) = \pi h(R - r)(R + r)$

98. $7921m^2 - 9409n^2 = (89m)^2 - (97n)^2 = (89m - 97n)(89m + 97n)$

Problem Set 2.7, pp. 86-87

2. $x^2 + 7x + 10$

$= (x + 2)(x + 5)$

4. $x^2 - 10x + 9$

$= (x - 1)(x - 9)$

6. $y^2 + 4y - 5$

$= (y + 5)(y - 1)$

8. $y^2 + 4y - 21$

$= (y + 7)(y - 3)$

10. $r^2 - 2r - 24$

$= (r - 6)(r + 4)$

12. $r^2 + 5r - 24$

$= (r + 8)(r - 3)$

14. $p^2 - 8p - 10$ is prime

16. $12 - 8p + p^2$

$= p^2 - 8p + 12$

$= (p - 2)(p - 6)$

18. $12t + t^2 - 13$

$= t^2 + 12t - 13$

$= (t + 13)(t - 1)$

20. $16 + 6t - t^2$

$= -t^2 + 6t + 16$

$= -(t^2 - 6t - 16)$

$= -(t - 8)(t + 2)$

22. $x^2 + 6xy + 5y^2$

$= (x + 5y)(x + y)$

24. $x^2 + 11xy - 12y^2$

$= (x + 12y)(x - y)$

26. $r^2 - 2rs - 35s^2$

$= (r - 7s)(r + 5s)$

28. $p^2 - 9pq + 20q^2$

$= (p - 4q)(p - 5q)$

30. $2x^2 + 7x + 3$

$= (2x + 1)(x + 3)$

32. $3x^2 - 8x + 5$

    $= (3x - 5)(x - 1)$

34. $2y^2 - 5y - 3$

    $= (2y + 1)(y - 3)$

36. $13y^2 + 2y - 11$

    $= (13y - 11)(y + 1)$

38. $3m^2 - 13m - 10$

    $= (3m + 2)(m - 5)$

40. $6x^2 - 13x - 5$

    $= (3x + 1)(2x - 5)$

42. $12r^2 - 17r - 5$

    $= (4r + 1)(3r - 5)$

44. $6t^2 + 5t - 10$ is prime

46. $9s^2 - 14s - 8$

    $= (9s + 4)(s - 2)$

48. $12 + 11x + 2x^2$

    $= 2x^2 + 11x + 12$

    $= (2x + 3)(x + 4)$

50. $15 - 2x^2 - 7x$

    $= -2x^2 - 7x + 15$

    $= -(2x^2 + 7x - 15)$

    $= -(2x - 3)(x + 5)$

52. $8p^2 - 27p - 20$

    $= (8p + 5)(p - 4)$

54. $3x^2 + 4xy + y^2$

    $= (3x + y)(x + y)$

56. $2x^2 - 5xy + 3y^2$

    $= (2x - 3y)(x - y)$

58. $4m^2 - 11mn - 3n^2$

    $= (4m + n)(m - 3n)$

60. $36p^2 + 5pq - 24q^2$

    $= (9p + 8q)(4p - 3q)$

62. $x^2 + 4x + 4$

    $= (x + 2)^2$

64. $z^2 - 8z + 16$

    $= (z - 4)^2$

66. $9t^2 + 6t + 1$

    $= (3t + 1)^2$

68. $9y^2 - 12y + 4$

    $= (3y - 2)^2$

70. $49 + 14r + r^2$

    $= r^2 + 14r + 49$

    $= (r + 7)^2$

72. $m^2 + 2mn + n^2$

    $= (m + n)^2$

74. $4r^2 - 20rs + 25s^2$

    $= (2r - 5s)^2$

76. $81s^2t^2 + 18st + 1$

= $(9st + 1)^2$

78. $3x^2 - 6x - 24$

= $3(x^2 - 2x - 8)$

= $3(x - 4)(x + 2)$

80. $7x^2 - 7x + 7$

= $7(x^2 - x + 1)$

82. $15x^2y + 180xy + 165y$

= $15y(x^2 + 12x + 11)$

= $15y(x + 1)(x + 11)$

84. $12m^3n^2 - 42m^2n^3 - 54mn^4$

= $6mn^2(2m^2 - 7mn - 9n^2)$

= $6mn^2(2m - 9n)(m + n)$

86. $z^4 + 7z^2 + 12$

= $(z^2 + 3)(z^2 + 4)$

88. $6t^4 + 5t^2 - 25$

= $(3t^2 - 5)(2t^2 + 5)$

90. $w^5 - 8w^3 - 9w$

= $w(w^4 - 8w^2 - 9)$

= $w(w^2 - 9)(w^2 + 1)$

= $w(w - 3)(w + 3)(w^2 + 1)$

92. $6r^4 + 25r^2s^2 - 25s^4$

= $(6r^2 - 5s^2)(r^2 + 5s^2)$

94.   a) $(2x + 4)^2 - 7(2x + 4) + 12$

= $4x^2 + 16x + 16 - 14x - 28 + 12$

= $4x^2 + 2x$

= $2x(2x + 1)$

b) $(2x + 4)^2 - 7(2x + 4) + 12$

= $u^2 - 7u + 12$ Let $2x + 4 = u$

= $(u - 3)(u - 4)$

= $((2x + 4) - 3)((2x + 4) - 4)$ Let $u = 2x + 4$

= $(2x + 1)2x$

96. $2x^2 - 5x - 3 = (2x + 1)(x - 3)$

a) $2(0.198)^2 - 5(0.198) - 3 \overset{?}{=} (2(0.198) + 1)(0.198 - 3)$

$$-3.911592 = -3.911592$$

b) $2(677)^2 - 5(677) - 3 \overset{?}{=} (2(677) + 1)(677 - 3)$

$$913,270 = 913,270$$

Problem Set 2.8, p. 92

2. $x^3 - 64 = x^3 - 4^3 = (x - 4)(x^2 + 4x + 16)$

4. $y^3 + 125 = y^3 + 5^3 = (y + 5)(y^2 - 5y + 25)$

6. $z^3 - 1 = z^3 - 1^3 = (z - 1)(z^2 + z + 1)$

8. $27 - r^3 = 3^3 - r^3 = (3 - r)(9 + 3r + r^2)$

10. $8x^3 + y^3 = (2x)^3 + y^3 = (2x + y)(4x^2 - 2xy + y^2)$

12. $125s^3 + 8t^3 = (5s)^3 + (2t)^3 = (5s + 2t)(25s^2 - 10st + 4t^2)$

14. $p^3q^3 - 1 = (pq)^3 - 1^3 = (pq - 1)(p^2q^2 + pq + 1)$

16. $125u^3 - 216v^3 = (5u)^3 - (6v)^3 = (5u - 6v)(25u^2 + 30uv + 36v^2)$

18. $x^6 + 27y^3 = (x^2)^3 + (3y)^3 = (x^2 + 3y)(x^4 - 3x^2y + 9y^2)$

20. $x^9 - 27 = (x^3)^3 - 3^3 = (x^3 - 3)(x^6 + 3x^3 + 9)$

22. $(m - 4)^3 + 8 = (m - 4)^3 + 2^3$

$$= ((m - 4) + 2)((m - 4)^2 - (m - 4)2 + 4)$$

$$= (m - 2)(m^2 - 8m + 16 - 2m + 8 + 4)$$

$$= (m - 2)(m^2 - 10m + 28)$$

24. $(p - q)^3 - (p + q)^3$

$$= ((p - q) - (p + q))((p - q)^2 + (p - q)(p + q) + (p + q)^2)$$

$$= (p - q - p - q)(p^2 - 2pq + q^2 + p^2 - q^2 + p^2 + 2pq + q^2)$$

$$= -2q(3p^2 + q^2)$$

26. $x^3 + 4x^2 + 5x + 20 = x^2(x + 4) + 5(x + 4) = (x + 4)(x^2 + 5)$

28. $ab + 11a + 3b + 33 = a(b + 11) + 3(b + 11) = (b + 11)(a + 3)$

30. $rs - r + 6s - 6 = r(s - 1) + 6(s - 1) = (s - 1)(r + 6)$

32. $3x^2 + 3xy + 7x + 7y = 3x(x + y) + 7(x + y) = (x + y)(3x + 7)$

34. $uv + v - u - 1 = v(u + 1) - 1(u + 1) = (u + 1)(v - 1)$

36. $s^3 + 9s^2t + 7st + 63t^2 = s^2(s + 9t) + 7t(s + 9t)$

$$= (s + 9t)(s^2 + 7t)$$

38. $a^2 + 4a + 4 - b^2 = (a + 2)^2 - b^2$

$$= ((a + 2) - b)((a + 2) + b)$$

$$= (a + 2 - b)(a + 2 + b)$$

40. $x^2 + 10x + 25 - 9y^2 = (x + 5)^2 - (3y)^2$

$$= ((x + 5) - 3y)((x + 5) + 3y)$$

$$= (x + 5 - 3y)(x + 5 + 3y)$$

42. $a^2 - b^2 + 8b - 16 = a^2 - (b^2 - 8b + 16)$

$$= a^2 - (b - 4)^2$$

$$= (a - (b - 4))(a + (b - 4))$$

$$= (a - b + 4)(a + b - 4)$$

44. $9r^2 - 12rs + 4s^2 - 25t^2 = (3r - 2s)^2 - (5t)^2$

$$= ((3r - 2s) - 5t)((3r - 2s) + 5t)$$

$$= (3r - 2s - 5t)(3r - 2s + 5t)$$

46. $x^2y - x^2 - y + 1 = x^2(y - 1) - 1(y - 1)$

$$= (y - 1)(x^2 - 1)$$

$$= (y - 1)(x - 1)(x + 1)$$

48. $3y^3 + y^2 - 18y - 6 = y^2(3y + 1) - 6(3y + 1)$

$$= (3y + 1)(y^2 - 6)$$

50. $p^2q^2 - 9q^2 + 4p^2 - 36 = q^2(p^2 - 9) + 4(p^2 - 9)$

$$= (p^2 - 9)(q^2 + 4)$$

$$= (p - 3)(p + 3)(q^2 + 4)$$

52. $t^4 - 2t^3 + 27t - 54 = t^3(t - 2) + 27(t - 2)$

$$= (t - 2)(t^3 + 27)$$

$$= (t - 2)(t + 3)(t^2 - 3t + 9)$$

54. $t^4 - s^3t - st^3 + s^4 = t(t^3 - s^3) - s(t^3 - s^3)$

$$= (t^3 - s^3)(t - s)$$

$$= (t - s)(t^2 + ts + s^2)(t - s)$$

$$= (t - s)^2(t^2 + ts + s^2)$$

56. $3x^3 + 24 = 3(x^3 + 8) = 3(x + 2)(x^2 - 2x + 4)$

58. $4000x^3 - 32 = 32(125x^3 - 1) = 32(5x - 1)(25x^2 + 5x + 1)$

60. $54r^3s^2 + 16s^2 = 2s^2(27r^3 + 8) = 2s^2(3r + 2)(9r^2 - 6r + 4)$

62. $15v^6 - 5v^3 - 10 = 5(3v^6 - v^3 - 2)$

$$= 5(3v^3 + 2)(v^3 - 1)$$

$$= 5(3v^3 + 2)(v - 1)(v^2 + v + 1)$$

64. $27x^{12n} - 64y^{6m} = (3x^{4n})^3 - (4y^{2m})^3$

$$= (3x^{4n} - 4y^{2m})(9x^{8n} + 12x^{4n}y^{2m} + 16y^{4m})$$

66. $k^6 - 729 = (k^3 - 27)(k^3 + 27)$

$$= (k - 3)(k^2 + 3k + 9)(k + 3)(k^2 - 3k + 9)$$

68. $(a - b)(a^2 + ab + b^2) = a^3 + a^2b + ab^2 - a^2b - ab^2 - b^3$

$$= a^3 - b^3$$

70. $19rs + 1007r + 29s + 1537 = 19r(s + 53) + 29(s + 53)$

$$= (s + 53)(19r + 29)$$

44

NOTES

CHAPTER 3

FIRST-DEGREE EQUATIONS AND INEQUALITIES

Problem Set 3.1, pp. 101-102

2. $2x + 3x + 1 = 21$

       $5x + 1 = 21$

          $5x = 20$

            $x = 4$

4. $3(x - 2) + 4 = 10$

    $3x - 6 + 4 = 10$

       $3x - 2 = 10$

          $3x = 12$

            $x = 4$

6. $3z + 5 = 2z + 12$

       $3z = 2z + 7$

         $z = 7$

8. $t + 9 = -t + 3$

       $t = -t - 6$

      $2t = -6$

       $t = -3$

10. $8m - 2 = 3m - 2$

       $8m = 3m$

       $5m = 0$

        $m = 0$

12. $5(y + 2) = y + 14$

    $5y + 10 = y + 14$

       $5y = y + 4$

       $4y = 4$

        $y = 1$

14. $5 - (p - 1) = 2p + 15$

    $5 - p + 1 = 2p + 15$

      $-p + 6 = 2p + 15$

         $-p = 2p + 9$

        $-3p = 9$

          $p = -3$

16. $6 + 4(2q + 5) = 13 - 15$

    $6 + 8q + 20 = -2$

       $8q + 26 = -2$

          $8q = -28$

            $q = -\dfrac{28}{8}$

            $q = -\dfrac{7}{2}$

18. $r - (2r + 3) = 5$

$r - 2r - 3 = 5$

$-r - 3 = 5$

$-r = 8$

$r = -8$

20. $x - (5x - 1) = 5x + 28$

$x - 5x + 1 = 5x + 28$

$-4x + 1 = 5x + 28$

$-4x = 5x + 27$

$-9x = 27$

$x = -3$

22. $3(r - 5) - 2(6r - 7) = -2$

$3r - 15 - 12r + 14 = -2$

$-9r - 1 = -2$

$-9r = -1$

$r = \dfrac{1}{9}$

24. $9 - 7(2s - 3) + 6s = 62$

$9 - 14s + 21 + 6s = 62$

$-8s + 30 = 62$

$-8s = 32$

$s = -4$

26. $t - (10t - 3) = -7 - 5(3t - 2)$

$t - 10t + 3 = -7 - 15t + 10$

$-9t + 3 = -15t + 3$

$-9t = -15t$

$6t = 0$

$t = 0$

28. $-(-4 + 6n) - (7 - 7n) - n = -(4 + 9n) + n + 1$

$4 - 6n - 7 + 7n - n = -4 - 9n + n + 1$

$-3 = -3 - 8n$

$0 = -8n$

$0 = n$

30. $11 - 3[r - 3(r - 1)] = 3 + 5r - 6[7 - (1 - r)]$

$\qquad 11 - 3[r - 3r + 3] = 3 + 5r - 6[7 - 1 + r]$

$\qquad\quad 11 - 3[-2r + 3] = 3 + 5r - 6[6 + r]$

$\qquad\qquad 11 + 6r - 9 = 3 + 5r - 36 - 6r$

$\qquad\qquad\qquad 6r + 2 = -r - 33$

$\qquad\qquad\qquad\quad 6r = -r - 35$

$\qquad\qquad\qquad\quad 7r = -35$

$\qquad\qquad\qquad\quad r = -5$

32. $\dfrac{x}{3} + \dfrac{x}{6} = 5$

$\quad 6\left(\dfrac{x}{3} + \dfrac{x}{6}\right) = 6(5)$

$\qquad\quad 2x + x = 30$

$\qquad\qquad 3x = 30$

$\qquad\qquad\; x = 10$

34. $\dfrac{a}{2} - \dfrac{a}{10} = -2$

$\quad 10\left(\dfrac{a}{2} - \dfrac{a}{10}\right) = 10(-2)$

$\qquad\quad 5a - a = -20$

$\qquad\qquad 4a = -20$

$\qquad\qquad\; a = -5$

36. $d - \dfrac{27}{8} = \dfrac{7d}{4} - \dfrac{7}{2}$

$\quad 8\left(d - \dfrac{27}{8}\right) = 8\left(\dfrac{7d}{4} - \dfrac{7}{2}\right)$

$\quad 8d - 27 = 14d - 28$

$\qquad\quad 8d = 14d - 1$

$\qquad\; -6d = -1$

$\qquad\qquad d = \dfrac{1}{6}$

38. $\dfrac{r}{2} - 2 = \dfrac{7r}{12} + \dfrac{2}{3}$

$\quad 12\left(\dfrac{r}{2} - 2\right) = 12\left(\dfrac{7r}{12} + \dfrac{2}{3}\right)$

$\quad 6r - 24 = 7r + 8$

$\qquad\quad 6r = 7r + 32$

$\qquad\; -r = 32$

$\qquad\quad r = -32$

40. $\dfrac{h + 1}{3} - \dfrac{h}{4} - \dfrac{1}{2} = 0$

$12\left(\dfrac{h + 1}{3} - \dfrac{h}{4} - \dfrac{1}{2}\right) = 12(0)$

$4(h + 1) - 3h - 6 = 0$

$4h + 4 - 3h - 6 = 0$

$h - 2 = 0$

$h = 2$

42. $\dfrac{k + 7}{2} - \dfrac{1}{6} = \dfrac{2}{3} - \dfrac{k + 9}{9}$

$18\left(\dfrac{k + 7}{2} - \dfrac{1}{6}\right) = 18\left(\dfrac{2}{3} - \dfrac{k + 9}{9}\right)$

$9(k + 7) - 3 = 12 - 2(k + 9)$

$9k + 63 - 3 = 12 - 2k - 18$

$9k + 60 = -2k - 6$

$9k = -2k - 66$

$11k = -66$

$k = -6$

.44. $\dfrac{2m - 3}{6} + 2 = 3 - \dfrac{m - 1}{9}$

$18\left(\dfrac{2m - 3}{6} + 2\right) = 18\left(3 - \dfrac{m - 1}{9}\right)$

$3(2m - 3) + 36 = 54 - 2(m - 1)$

$6m - 9 + 36 = 54 - 2m + 2$

$6m + 27 = -2m + 56$

$6m = -2m + 29$

$8m = 29$

$m = \dfrac{29}{8}$

46. $6.2y - 1.1 = 8.9y + 7$

$6.2y = 8.9y + 8.1$

$-2.7y = 8.1$

$y = -3$

48. $0.35q + 0.05(6 - q) = 2.4(2)$

$0.35q + 0.3 - 0.05q = 4.8$

$0.3q + 0.3 = 4.8$

$0.3q = 4.5$

$q = 15$

50. $4(x - 2) = 3x + 4$

$4x - 8 = 3x + 4$

$4x = 3x + 12$

$x = 12$

Conditional equation

52. $4(x - 2) = 4x - 8$

$4x - 8 = 4x - 8$

Identity

54. $4(x - 2) = 4(x + 1)$

$4x - 8 = 4x + 4$

$-8 = 4$

Contradiction

56. $6x - (9x - 3) = 4 + 3(1 - x)$

$6x - 9x + 3 = 4 + 3 - 3x$

$-3x + 3 = 7 - 3x$

$3 = 7$

Contradiction

58. $\frac{y + 3}{9} - \frac{y}{6} = \frac{1}{3} - \frac{y}{18}$

$18(\frac{y + 3}{9} - \frac{y}{6}) = 18(\frac{1}{3} - \frac{y}{18})$

$2(y + 3) - 3y = 6 - y$

$2y + 6 - 3y = 6 - y$

$6 - y = 6 - y$

Identity

60. $x(x - 6) - x(x - 3) = 12$

$x^2 - 6x - x^2 + 3x = 12$

$-3x = 12$

$x = -4$

62. $2u(2u - 3) + 3u = 4u^2 + 42$

$4u^2 - 6u + 3u = 4u^2 + 42$

$-3u = 42$

$u = -14$

64.

$$(w + 5)^2 - (w + 8)^2 = 1$$

$(w^2 + 10w + 25) - (w^2 + 16w + 64) = 1$

$w^2 + 10w + 25 - w^2 - 16w - 64 = 1$

$-6w - 39 = 1$

$-6w = 40$

$w = -\frac{20}{3}$

66. $v - (v + 1)^2 = -v^2 - 9$ 68. $2n = 5 + 6n$

$v - (v^2 + 2v + 1) = -v^2 - 9$ $-4n = 5$

$v - v^2 - 2v - 1 = -v^2 - 9$ $n = -\dfrac{5}{4}$

$-v^2 - v - 1 = -v^2 - 9$

$-v - 1 = -9$

$-v = -8$

$v = 8$

70. $p - 0.20p = 68$ 72. $\dfrac{79 + 86 + 90 + 82 + 3x}{7} = 85$

$0.8p = 68$ $\dfrac{337 + 3x}{7} = 85$

$p = \$85$ $337 + 3x = 595$

$3x = 258$

$x = 86$

74. $q = \dfrac{q}{353} - \dfrac{839}{797}$

$281,341q = 281,341\left(\dfrac{q}{353} - \dfrac{839}{797}\right)$

$281,341q = 797q - 296,167$

$280,544q = -296,167$

$q \approx -1.06$

Problem Set 3.2, pp. 106-109

2. $A = \ell w$ 4. $A = \dfrac{1}{2}bh$

$30 = \ell(5)$ $18 = \dfrac{1}{2}(9)h$

$6 = \ell$ $36 = 9h$

$4 = h$

6. $V = P + Pr$

   $1250 = 1100 + 1100r$

   $150 = 1100r$

   $\dfrac{150}{1100} = r$

   $\dfrac{3}{22} = r$

8. $S = \dfrac{a}{1 - r}$

   $48 = \dfrac{a}{1 - (-1/3)}$

   $48 = \dfrac{a}{4/3}$

   $\dfrac{4}{3} \cdot 48 = \dfrac{a}{4/3} \cdot \dfrac{4}{3}$

   $64 = a$

10. $T = 2\pi r(h + r)$

   $109.9 = 2(3.14)3.5(h + 3.5)$

   $109.9 = 21.98(h + 3.5)$

   $5 = h + 3.5$

   $1.5 = h$

12. Method I: $d = rt$

   $165 = 55t$

   $3 \text{ hr} = t$

   Method II: $t = \dfrac{d}{r}$

   $t = \dfrac{165}{55} = 3 \text{ hr}$

14. $P = 2w + 2\ell$

   $P - 2\ell = 2w$

   $\dfrac{P - 2\ell}{2} = w$

   $\dfrac{46 - 2(15)}{2} = w$

   $\dfrac{16}{2} = w$

   $8 \text{ m} = w$

16. $F = \dfrac{9}{5}C + 32$

   $F - 32 = \dfrac{9}{5}C$

   $\dfrac{5}{9}(F - 32) = C$

   $\dfrac{5}{9}(5 - 32) = C$

   $\dfrac{5}{9}(-27) = C$

   $-15° = C$

18. $y - 4x + 16 = 0$

   $y = 4x - 16$

20. $V = \dfrac{1}{2}bh$

   $2V = bh$

   $\dfrac{2V}{h} = b$

22. $s = \frac{1}{2}(a + b + c)$ 24. $5y - x + 10 = 0$

$2s = a + b + c$ $5y = x - 10$

$2s - a - c = b$ $y = \dfrac{x - 10}{5}$

26. $V = P + Prt$ 28. $V = P + Prt$

$V - P = Prt$ $V = P(1 + rt)$

$\dfrac{V - P}{Pr} = t$ $\dfrac{V}{1 + rt} = P$

30. $s = vt + \frac{1}{3}at^2$ 32. $4x - 3y + 1 = 0$

$s - vt = \frac{1}{3}at^2$ $4x + 1 = 3y$

$3(s - vt) = at^2$ $\dfrac{4x + 1}{3} = y$

$\dfrac{3(s - vt)}{t^2} = a$

34. $c^2 = a^2 + b^2$ 36. $S = 4\pi r^2$

$c^2 - b^2 = a^2$ $\dfrac{S}{4\pi} = r^2$

38. $B = 2b + (n - 2)d$ 40. $T = 2\pi r(h + r)$

$B - 2b = (n - 2)d$ $\dfrac{T}{2\pi r} = h + r$

$\dfrac{B - 2b}{d} = n - 2$

$\dfrac{B - 2b}{d} + 2 = n$ $\dfrac{T}{2\pi r} - r = h$

42. $a(x + b) = 5 + 2x$

$ax + ab = 5 + 2x$

$ax - 2x = 5 - ab$

$x(a - 2) = 5 - ab$

$x = \dfrac{5 - ab}{a - 2}$

44. A = area of shaded region

$$A = \begin{matrix} \text{area of} \\ \text{rectangle} \end{matrix} - [\begin{matrix} \text{area of} \\ \text{triangle} \end{matrix} + \begin{matrix} \text{one half area} \\ \text{of circle} \end{matrix}]$$

$$A = \ell w - [\tfrac{1}{2}bh + \tfrac{1}{2}\pi r^2]$$

$$A = 8(14) - [\tfrac{1}{2}(8)(10) + \tfrac{1}{2}(3.14)(4^2)]$$

$$A = 112 - [40 + 25.12]$$

$$A \approx 46.9 \text{ cm}^2$$

46. A = total area

$$A = \begin{matrix} \text{area of} \\ \text{floor} \end{matrix} + \begin{matrix} \text{area of} \\ \text{ceiling} \end{matrix} + \begin{matrix} \text{area of} \\ \text{four walls} \end{matrix}$$

$$A = s^2 + s^2 + 4s^2$$

$$A = 6s^2$$

$$A = 6(12^2)$$

$$A = 864 \text{ ft}^2$$

48. V = volume of cylinder

$$V = \pi r^2 h$$

$$V = 3.14(5^2)(6)$$

$$V = 471 \text{ ft}^3$$

50. d = rt

$$d = 1100(0.3)$$

$$d = 330 \text{ ft}$$

52. To average 60 mph over 2 miles the car must travel the
     2 miles in

$$t = \frac{d}{r} = \frac{2}{60} = \frac{1}{30} \text{ hr} = 2 \text{ min.}$$

However the car used 2 minutes to travel the first mile at
30 mph, because

$$t = \frac{d}{r} = \frac{1}{30} \text{ hr} = 2 \text{ min.}$$

Therefore it cannot be done.

54.

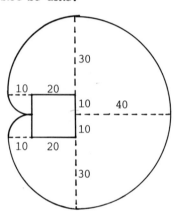

A = grazing area

      one half area   one half area   one half area
A = of circle with + of circle with + of circle with
      radius 40   radius 30   radius 10

$$A = \frac{1}{2}(3.14)(40^2) + \frac{1}{2}(3.14)(30^2) + \frac{1}{2}(3.14)(10^2)$$

$$A = 2512 + 1413 + 157$$

$$A \approx 4080 \text{ ft}^2$$

56. $F = \frac{9}{5}C + 32$

   $C = \frac{9}{5}C + 32$ Let $F = C$

   $5C = 9C + 160$

   $-4C = 160$

   $C = -40°$

58.

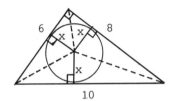

       10

Area of large triangle = sum of areas of three small triangles

$$\frac{1}{2}(6)(8) = \frac{1}{2}(10)x + \frac{1}{2}(6)x + \frac{1}{2}(8)x$$

$$24 = 5x + 3x + 4x$$

$$24 = 12x$$

$$2 \text{ cm} = x$$

60.   385 yd = 1155 ft = $\frac{1155}{5280}$ mi = 0.21875 mi

   7 min 11 sec = 431 sec = $\frac{431}{3600}$ hr $\approx$ 0.11972 hr

   $r = \frac{d}{t} \approx \frac{26.21875 \text{ mi}}{2.11972 \text{ hr}} \approx 12.4$ mph

Problem Set 3.3, pp. 112-114

2.   x = smaller number

4x = larger number

sum = 65

x + 4x = 65

5x = 65

x = 13

4x = 52

4.   x = first integer

x + 2 = second integer

sum = 256

x + x + 2 = 256

2x + 2 = 256

2x = 254

x = 127

x + 2 = 129

6.   x = first integer

x + 1 = second integer

x + 2 = third integer

x + 3 = fourth integer

$\underline{\text{One fifth}}$ $\underline{\text{the sum}}$ = 18

$\frac{1}{5}(x + x + 1 + x + 2 + x + 3)$ = 18

$\frac{1}{5}(4x + 6)$ = 18

4x + 6 = 90

4x = 84

x = 21

x + 1 = 22

x + 2 = 23

x + 3 = 24

8.   x = first integer

x + 1 = second integer

x + 2 = third integer

x + 3 = fourth integer

$\underset{\text{the first}}{\text{Three times}}$ $-$ $\underset{\text{third}}{\text{the}}$ $=$ $\underset{\text{the fourth}}{\text{12 more tha}}$

3x $-$ (x + 2) = 12 + (x + 3

3x $-$ x $-$ 2 = 15 + x

2x $-$ 2 = 15 + x

2x = 17 + x

x = 17

x + 1 = 18

x + 2 = 19

x + 3 = 20

10.   x = first integer

   x + 2 = second integer

   x + 4 = third integer

$$\underset{\text{the largest}}{\text{Three times}} - \underset{\text{smallest}}{\text{the}} = 50$$

$$3(x + 4) - x = 50$$

$$3x + 12 - x = 50$$

$$2x + 12 = 50$$

$$2x = 38$$

$$x = 19$$

Since x must be even, there
is no solution.

12.   w = width

   2w + 24 = length

$$\underset{\text{width}}{\text{Twice the}} + \underset{\text{length}}{\text{twice the}} = \underset{\text{perimeter}}{\text{the}}$$

$$2w + 2(2w + 24) = 210$$

$$2w + 4w + 48 = 210$$

$$6w + 48 = 210$$

$$6w = 162$$

$$w = 27 \text{ ft}$$

$$2w + 24 = 78 \text{ ft}$$

14.   x = first angle

   3x − 4 = second angle

   3(3x − 4) − 38 = third angle

$$\text{Sum of angles} = 180$$

$$x + 3x - 4 + 3(3x - 4) - 38 = 180$$

$$x + 3x - 4 + 9x - 12 - 38 = 180$$

$$13x - 54 = 180$$

$$13x = 234$$

$$x = 18°$$

$$3x - 4 = 50°$$

$$3(3x - 4) - 38 = 112°$$

16. w = width

   $w + 4$ = length

New area = old area + 40

$(w + 2)(w + 6) = w(w + 4) + 40$

$w^2 + 8w + 12 = w^2 + 4w + 40$

$8w + 12 = 4w + 40$

$8w = 4w + 28$

$4w = 28$

$w = 7$ m

$w + 4 = 11$ m

18. y = Jimmy's age

$y + 8$ = Jerry's age

$\dfrac{\text{Jerry's age}}{\text{in 2 yr}} = \dfrac{\text{twice Jimmy's}}{\text{age in 2 yr}}$

$y + 8 + 2 = 2(y + 2)$

$y + 10 = 2y + 4$

$y = 2y - 6$

$-y = -6$

$y = 6$ yr

$y + 8 = 14$ yr

20. y = Eric's age

$3y$ = brother's age

$\dfrac{\text{Eric's age}}{\text{in 9 yr}} = \dfrac{\text{brother's age}}{\text{13 yr ago}}$

$y + 9 = 3y - 13$

$y = 3y - 22$

$-2y = -22$

$y = 11$ yr

$3y = 33$ yr

22.　　　 n = no. of dimes

　　 n + 3 = no. of quarters

　　　 2n = no. of nickels

$$\underset{\text{dimes}}{\text{Value of}} + \underset{\text{quarters}}{\text{value of}} + \underset{\text{nickels}}{\text{value of}} = \underset{\text{value}}{\text{total}}$$

$$10n + 25(n+3) + 5(2n) = 435$$

$$10n + 25n + 75 + 10n = 435$$

$$45n + 75 = 435$$

$$45n = 360$$

$$n = 8 \text{ dimes}$$

$$n + 3 = 11 \text{ quarters}$$

$$2n = 16 \text{ nickels}$$

24.　　　 n = no. of dimes

　 60 - n = no. of quarters

$$\underset{\text{dimes}}{\text{Value of}} + \underset{\text{quarters}}{\text{value of}} = \underset{\text{value}}{\text{total}}$$

$$10n + 25(60-n) = 1065$$

$$10n + 1500 - 25n = 1065$$

$$1500 - 15n = 1065$$

$$-15n = -435$$

$$n = 29 \text{ dimes}$$

$$60 - n = 31 \text{ quarters}$$

26.   x = no. of correct answers

75 − x = no. of incorrect answers

$$\underset{\text{answers}}{\text{Correct}} - \underset{\text{incorrect answers}}{\text{one fourth}} = \text{score}$$

$$x - \frac{1}{4}(75 - x) = 65$$

$$x - \frac{75}{4} + \frac{1}{4}x = 65$$

$$-\frac{75}{4} + \frac{5}{4}x = 65$$

$$-75 + 5x = 260$$

$$5x = 335$$

$$x = 67 \text{ correct}$$

$$75 - x = 8 \text{ incorrect}$$

28. x = the number

$$\underset{\text{by a 1}}{\text{Number followed}} = \underset{\text{with 1 in front}}{\text{three times number}}$$

$$10x + 1 = 3(x + 100{,}000)$$

$$10x + 1 = 3x + 300{,}000$$

$$10x = 3x + 299{,}999$$

$$7x = 299{,}999$$

$$x = 42{,}857$$

30. Since Mr. Smith's tax was $18,384.76, his taxable income must have been between $42,800 and $54,700.

x = Mr. Smith's taxable income over $42,800

$$13{,}639 + 0.48x = 18{,}384.76$$

$$0.48x = 4745.76$$

$$x = 9887$$

Mr. Smith's total taxable income = 42,800 + 9887 = $52,687

Problem Set 3.4, pp. 117–119

2.   x = amount at 9%

   8800 − x = amount at 13%

   Interest on = interest on
   9% investment   13% investment

         $0.09x = 0.13(8800 - x)$

         $0.09x = 1144 - 0.13x$

         $0.22x = 1144$

             $x = \$5200$

     $8800 - x = \$3600$

4. x = amount at 9%

   Interest on + interest on = 500
   9% investment   8% investment

      $0.09x \quad + \quad 0.08(4000) \quad = 500$

                $0.09x + 320 = 500$

                     $0.09x = 180$

                         $x = \$2000$

6.   x = no. of reserved seats

   700 − x = no. of general admission seats

   Money from + money from general = total
   reserved seats   admission seats   receipts

      $4.50x \quad + \quad 2.50(700 - x) \quad = 2390$

                $4.5x + 1750 - 2.5x = 2390$

                      $1750 + 2x = 2390$

                            $2x = 640$

                             $x = 320$ reserved

                   $700 - x = 380$ general admission

8.   x = no. of grams of 15% alloy

     25 - x = no. of grams of 40% alloy

     Pure silver   + pure silver   = pure silver
     in 15% alloy   in 40% alloy   in 20% alloy
          ↓   ↓   ↓
        0.15x   + 0.40(25 - x) = 0.20(25)

            0.15x + 10 - 0.40x = 5

                 10 - 0.25x = 5

                    -0.25x = -5

                        x = 20 g

                   25 - x = 5 g

10. x = no. of liters of pure (100%) formaldehyde

     Pure formaldehyde + pure formaldehyde = pure formaldehyde
     in 100% solution   in 40% solution   in 50% solution
           ↓   ↓   ↓
         1.00x   +   0.40(15)   = 0.50(x + 15)

                          x + 6 = 0.5x + 7.5

                              x = 0.5x + 1.5

                           0.5x = 1.5

                              x = 3 ℓ

12. x = no. of quarts drained and replaced with pure (100%)
         antifreeze

     Pure antifreeze in + pure antifreeze in = pure antifreeze in
     replaced solution   solution not replaced   final solution
           ↓   ↓   ↓
         1.00x   +   0.20(12 - x)   = 0.50(12)

                       x + 2.4 - 0.2x = 6

                          2.4 + 0.8x = 6

                              0.8x = 3.6

                                 x = 4.5 qt

14.   t = time P-wave travels

t + 45 = time S-wave travels

Distance of P-wave = distance of S-wave

$$5t = 2.75(t + 45)$$

$$5t = 2.75t + 123.75$$

$$2.25t = 123.75$$

$$t = 55 \text{ sec}$$

Distance to epicenter = 5(55) = 275 mi

16.   t = time second jogger travels

$t + \dfrac{1}{4}$ = time first jogger travels

Distance of first jogger = distance of second jogger

$$10\left(t + \frac{1}{4}\right) = 12t$$

$$10t + 2.5 = 12t$$

$$2.5 = 2t$$

$$1.25 \text{ hr} = t$$

18. t = time for fast runner to lap slow runner

$$\text{Distance of fast runner} = \text{distance of slow runner} + 1 \text{ mile}$$

$$12t = 8t + 1$$

$$4t = 1$$

$$t = \frac{1}{4} \text{ hr}$$

20.   r = rate of slow runner

r + 3 = rate of fast runner

Distance of slow runner = distance of fast runner

$$r(1\tfrac{1}{3}) = (r + 3)\cdot 1$$

$$\tfrac{4}{3}r = r + 3$$

$$\tfrac{1}{3}r = 3$$

$$r = 9 \text{ mph}$$

$$r + 3 = 12 \text{ mph}$$

22. d = distance upstream

$$\dfrac{\text{Time}}{\text{upstream}} + \dfrac{\text{time}}{\text{downstream}} = \dfrac{\text{total}}{\text{time}}$$

$$\dfrac{d}{8} + \dfrac{d}{12} = 6$$

$$24\left(\dfrac{d}{8} + \dfrac{d}{12}\right) = 24(6)$$

$$3d + 2d = 144$$

$$5d = 144$$

$$d = 28.8 \text{ mi}$$

24. x = no. of calculators

Revenue = cost

$$18.50x = 6x + 1400$$

$$12.5x = 1400$$

$$x = 112 \text{ calculators}$$

26. x = no. of additional seats

$$\dfrac{\text{Revenue from}}{\text{new seating}} = \dfrac{\text{revenue from}}{\text{old seating}}$$

$$6(15,000 + x) = 6.50(15,000)$$

$$90,000 + 6x = 97,500$$

$$6x = 7500$$

$$x = 1250 \text{ seats}$$

28. t = time to stuff 1656 envelopes

$$\underset{\text{by Tom}}{\text{Envelopes stuffed}} + \underset{\text{by Lisa}}{\text{envelopes stuffed}} = 1656$$

$$32t \qquad + \qquad 40t \qquad = 1656$$

$$72t = 1656$$

$$t = 23 \text{ min}$$

30. d = distance to target

$$\underset{\text{to hit target}}{\text{Time for bullet}} + \underset{\text{to reach policeman}}{\text{time for sound}} = 1.5$$

$$\frac{d}{773} \qquad + \qquad \frac{d}{1100} \qquad = 1.5$$

$$(773)(1100)(\frac{d}{773} + \frac{d}{1100}) = (773)(1100)(1.5)$$

$$1100d + 773d = 1,275,450$$

$$1873d = 1,275,450$$

$$d \approx 681 \text{ ft}$$

Problem Set 3.5, pp. 124-126

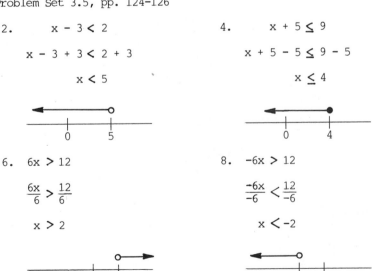

2. $x - 3 < 2$

$x - 3 + 3 < 2 + 3$

$x < 5$

4. $x + 5 \leq 9$

$x + 5 - 5 \leq 9 - 5$

$x \leq 4$

6. $6x > 12$

$\frac{6x}{6} > \frac{12}{6}$

$x > 2$

8. $-6x > 12$

$\frac{-6x}{-6} < \frac{12}{-6}$

$x < -2$

10. $-x < 4$

$\dfrac{-x}{-1} > \dfrac{4}{-1}$

$x > -4$

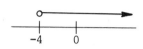

12. $2x + 5 < 11$

$2x + 5 - 5 < 11 - 5$

$2x < 6$

$\dfrac{2x}{2} < \dfrac{6}{2}$

$x < 3$

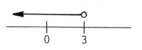

14. $-3y - 4 \leq 2$

$-3y - 4 + 4 \leq 2 + 4$

$-3y \leq 6$

$\dfrac{-3y}{-3} \geq \dfrac{6}{-3}$

$y \geq -2$

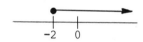

16. $-t + 17 \geq 16$

$-t + 17 - 17 \geq 16 - 17$

$-t \geq -1$

$\dfrac{-t}{-1} \leq \dfrac{-1}{-1}$

$t \leq 1$

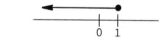

18. $-4m \leq 0$

$\dfrac{-4m}{-4} \geq \dfrac{0}{-4}$

$m \geq 0$

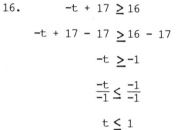

20. $6x - 8 < x + 2$

$6x < x + 10$

$5x < 10$

$x < 2$

22. $10 - 3y \leq 7y - 10$

$-3y \leq 7y - 20$

$-10y \leq -20$

$y \geq 2$

24. $5z + 1 \geq -2z + 1$

$5z \geq -2z$

$7z \geq 0$

$z \geq 0$

26. $2(m + 2) - 5(2m - 4) < 0$

$2m + 4 - 10m + 20 < 0$

$-8m + 24 < 0$

$-8m < -24$

$m > 3$

28. $5 - 2(r + 1) > 2r + 13$

$5 - 2r - 2 > 2r + 13$

$-2r + 3 > 2r + 13$

$-2r > 2r + 10$

$-4r > 10$

$r < -\dfrac{10}{4}$

$r < -\dfrac{5}{2}$

30. $3p - (5p + 14) < p + 4$

$3p - 5p - 14 < p + 4$

$-2p - 14 < p + 4$

$-2p < p + 18$

$-3p < 18$

$p > -6$

32. $2(2q - 1) \le 2 - (q + 4)$

$4q - 2 \le 2 - q - 4$

$4q - 2 \le -q - 2$

$4q \le -q$

$5q \le 0$

$q \le 0$

34. $\dfrac{3}{4}x \le -3$

$\dfrac{4}{3}(\dfrac{3}{4}x) \le \dfrac{4}{3}(-3)$

$x \le -4$

36. $\dfrac{y}{3} - 2 \ge y + \dfrac{2}{3}$

$3(\dfrac{y}{3} - 2) \ge 3(y + \dfrac{2}{3})$

$y - 6 \ge 3y + 2$

$y \ge 3y + 8$

$-2y \ge 8$

$y \le -4$

38. $\frac{1}{9}t + \frac{5}{18} > \frac{1}{6}t + \frac{2}{3}$

$18\left(\frac{1}{9}t + \frac{5}{18}\right) > 18\left(\frac{1}{6}t + \frac{2}{3}\right)$

$2t + 5 > 3t + 12$

$2t > 3t + 7$

$-t > 7$

$t < -7$

40. $\frac{3v - 1}{4} + 6 \leq 2 + \frac{v}{3}$

$12\left(\frac{3v - 1}{4} + 6\right) \leq 12\left(2 + \frac{v}{3}\right)$

$3(3v - 1) + 72 \leq 24 + 4v$

$9v - 3 + 72 \leq 24 + 4v$

$9v + 69 \leq 24 + 4v$

$9v \leq -45 + 4v$

$5v \leq -45$

$v \leq -9$

42. $1 < 10 - \frac{4u + 5}{5}$

$5(1) < 5\left(10 - \frac{4u + 5}{5}\right)$

$5 < 50 - (4u + 5)$

$5 < 50 - 4u - 5$

$5 < 45 - 4u$

$-40 < -4u$

$10 > u$

44. $\frac{3m}{7} - \frac{m - 4}{3} \geq 4 + \frac{2m}{7}$

$21\left(\frac{3m}{7} - \frac{m - 4}{3}\right) \geq 21\left(4 + \frac{2m}{7}\right)$

$9m - 7(m - 4) \geq 84 + 6m$

$9m - 7m + 28 \geq 84 + 6m$

$2m + 28 \geq 84 + 6m$

$2m \geq 56 + 6m$

$-4m \geq 56$

$m \leq -14$

46. $3 < x - 1 < 5$

$3 + 1 < x - 1 + 1 < 5 + 1$

$4 < x < 6$

48. $-4 \leq 2r \leq 10$

$\frac{-4}{2} \leq \frac{2r}{2} \leq \frac{10}{2}$

$-2 \leq r \leq 5$

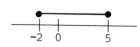

50. $-10 \leq 5s - 10 < 0$

$-10 + 10 \leq 5s - 10 + 10 < 0 + 10$

$0 \leq 5s < 10$

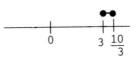

$\dfrac{0}{5} \leq \dfrac{5s}{5} < \dfrac{10}{5}$

$0 \leq s < 2$

52. $5 < 3 - 2y \leq 10$

$5 - 3 < 3 - 2y - 3 \leq 10 - 3$

$2 < -2y \leq 7$

$\dfrac{2}{-2} > \dfrac{-2y}{-2} \geq \dfrac{7}{-2}$

$-1 > y \geq -\dfrac{7}{2}$ or $-\dfrac{7}{2} \leq y < -1$

54. $-\dfrac{3}{2} \leq \dfrac{1 - 3q}{6} \leq -\dfrac{4}{3}$

$6\left(-\dfrac{3}{2}\right) \leq 6\left(\dfrac{1 - 3q}{6}\right) \leq 6\left(-\dfrac{4}{3}\right)$

$-9 \leq 1 - 3q \leq -8$

$-9 - 1 \leq 1 - 3q - 1 \leq -8 - 1$

$-10 \leq -3q \leq -9$

$\dfrac{-10}{-3} \geq \dfrac{-3q}{-3} \geq \dfrac{-9}{-3}$

$\dfrac{10}{3} \geq q \geq 3$ or $3 \leq q \leq \dfrac{10}{3}$

56. $t - 6 < 4t - 3 < t + 9$

$t - 6 - t < 4t - 3 - t < t + 9 - t$

$-6 < 3t - 3 < 9$

$-6 + 3 < 3t - 3 + 3 < 9 + 3$

$-3 < 3t < 12$

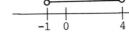

$\dfrac{-3}{3} < \dfrac{3t}{3} < \dfrac{12}{3}$

$-1 < t < 4$

58. $4(x + 1) - 5 < 11 + 4x$

$4x + 4 - 5 < 11 + 4x$

$4x - 1 < 11 + 4x$

$-1 < 11$

This statement is always true, so the inequality is an identity.

60. $6x - 3(x - 1) < 9 - (x - 2)$

$6x - 3x + 3 < 9 - x + 2$

$3x + 3 < -x + 11$

$3x < -x + 8$

$4x < 8$

$x < 2$

This statement is true when $x < 2$ and false when $x \geq 2$, so the inequality is conditional.

62. $\dfrac{3}{4} + \dfrac{12x - 1}{8} < \dfrac{3x}{2}$

$8\left(\dfrac{3}{4} + \dfrac{12x - 1}{8}\right) < 8\left(\dfrac{3x}{2}\right)$

$6 + 12x - 1 < 12x$

$12x + 5 < 12x$

$5 < 0$

This statement is always false, so the inequality is a contradiction.

64. $-9 < 6 + 5x < 21$

$-15 < 5x < 15$

$-3 < x < 3$

66. $23 < \dfrac{9}{5}C + 32 < 68$

$-9 < \dfrac{9}{5}C < 36$

$\dfrac{5}{9}(-9) < \dfrac{5}{9}\left(\dfrac{9}{5}C\right) < \dfrac{5}{9}(36)$

$-5° < C < 20$

68.   x = score on fifth test

$80 \leq \dfrac{95 + 87 + 80 + 99 + x}{5} < 90$

$80 \leq \dfrac{361 + x}{5} < 90$

$400 \leq 361 + x < 450$

$39 \leq x < 89$

70. x = no. of T-shirts produced and sold each day

Profit = revenue - cost

Profit = 7.20x - (5.60x + 1600) = 1.6x - 1600

a) 1.6x - 1600 > 0

$$1.6x > 1600$$

$$x > 1000 \text{ T-shirts}$$

b) 1.6x - 1600 > 400

$$1.6x > 2000$$

$$x > 1250 \text{ T-shirts}$$

c) 1.6x - 1600 > 0.20(5.60x + 1600)

$$1.6x - 1600 > 1.12x + 320$$

$$1.6x > 1.12x + 1920$$

$$0.48x > 1920$$

$$x > 4000 \text{ T-shirts}$$

72.   price < 40

500 - 1.25x < 40

$$- 1.25x < -460$$

$$x > 368 \text{ units}$$

Problem Set 3.6, pp. 131-132

2. $|x| = 2$

x = 2 or x = -2

4. $|2x| = 8$

2x = 8 or 2x = -8

x = 4   x = -4

6. $|y - 3| = 5$

y - 3 = 5 or y - 3 = -5

y = 8   y = -2

8. $|2x - 7| = 3$

2x - 7 = 3 or 2x - 7 = -3

2x = 10   2x = 4

x = 5   x = 2

10. $|4r| + 5 = 17$

$\qquad |4r| = 12$

$\qquad 4r = 12$ or $4r = -12$

$\qquad r = 3 \qquad r = -3$

12. $|5p + 6| - 3 = 18$

$\qquad |5p + 6| = 21$

$\qquad 5p + 6 = 21$ or $5p + 6 = -21$

$\qquad 5p = 15 \qquad 5p = -27$

$\qquad p = 3 \qquad p = -\dfrac{27}{5}$

14. $|8y + 3| = -5$

No solution because absolute value cannot be negative.

16. $|3x + 4| = |2x + 11|$

$\qquad 3x + 4 = 2x + 11$ or $3x + 4 = -(2x + 11)$

$\qquad 3x = 2x + 7 \qquad 3x + 4 = -2x - 11$

$\qquad x = 7 \qquad\qquad 3x = -2x - 15$

$\qquad\qquad\qquad\qquad 5x = -15$

$\qquad\qquad\qquad\qquad x = -3$

18. $|5t - 4| = |2t + 5|$

$\qquad 5t - 4 = 2t + 5$ or $5t - 4 = -(2t + 5)$

$\qquad 5t = 2t + 9 \qquad 5t - 4 = -2t - 5$

$\qquad 3t = 9 \qquad\qquad 5t = -2t - 1$

$\qquad t = 3 \qquad\qquad 7t = -1$

$\qquad\qquad\qquad\qquad t = -\dfrac{1}{7}$

20. $|4y - 9| = |3 - 4y|$

$\qquad 4y - 9 = 3 - 4y$ or $4y - 9 = -(3 - 4y)$

$\qquad 4y = 12 - 4y \qquad 4y - 9 = -3 + 4y$

$\qquad 8y = 12 \qquad\qquad 4y = 6 + 4y$

$\qquad y = \dfrac{12}{8} \qquad\qquad 0 = 6$

$\qquad y = \dfrac{3}{2} \qquad\qquad$ No solution

22. $|6v + 1| = |6v - 1|$

$6v + 1 = 6v - 1$ or $6v + 1 = -(6v - 1)$

$1 = -1$ $6v + 1 = -6v + 1$

No solution $6v = -6v$

                 $12v = 0$

                 $v = 0$

24. $|x| < 2$

$-2 < x < 2$

26. $|3x| < 6$

$-6 < 3x < 6$

$-2 < x < 2$

28. $|y - 2| < 5$

$-5 < y - 2 < 5$

$-3 < y < 7$

30. $|2z + 5| \leq 3$

$-3 \leq 2z + 5 \leq 3$

$-8 \leq 2z \leq -2$

$-4 \leq z \leq -1$

32. $|6k| - 12 \leq 0$

$|6k| \leq 12$

$-12 \leq 6k \leq 12$

$-2 \leq k \leq 2$

34. $4|2r + 5| - 33 < -5$

$4|2r + 5| < 28$

$|2r + 5| < 7$

$-7 < 2r + 5 < 7$

$-12 < 2r < 2$

$-6 < r < 1$

36. $\left|\dfrac{8s + 1}{3}\right| \leq 5$

$-5 \leq \dfrac{8s + 1}{3} \leq 5$

$-15 \leq 8s + 1 \leq 15$

$-16 \leq 8s \leq 14$

$-2 \leq s \leq \dfrac{7}{4}$

38. $|x| > 2$

$x < -2$ or $x > 2$

40. $|4x| > 12$

$4x < -12$ or $4x > 12$

$x < -3$ $x > 3$

42. $|r + 1| > 4$

$r + 1 < -4$ or $r + 1 > 4$

$r < -5$ $r > 3$

44. $|5s - 3| \geq 7$

$5s - 3 \leq -7$ or $5s - 3 \geq 7$

$5s \leq -4$ $5s \geq 10$

$s \leq -\frac{4}{5}$ $s \geq 2$

46. $11 \leq |3 - 4t|$

$|3 - 4t| \geq 11$

$3 - 4t \leq -11$ or $3 - 4t \geq 11$

$-4t \leq -14$ $-4t \geq 8$

$t \geq \frac{7}{2}$ $t \leq -2$

48. $\left|\frac{4}{5}t\right| - 8 > 0$

$\left|\frac{4}{5}t\right| > 8$

$\frac{4}{5}t < -8$ or $\frac{4}{5}t > 8$

$\frac{5}{4}(\frac{4}{5}t) < \frac{5}{4}(-8)$ $\frac{5}{4}(\frac{4}{5}t) > \frac{5}{4}(8)$

$t < -10$ $t > 10$

50. $|2w - 1| \geq -3$

Any real number is a
solution since absolute
value is always nonnegative.

52. $|3x - 4| = 8$

$3x - 4 = 8$ or $3x - 4 = -8$

$3x = 12$ $3x = -4$

$x = 4$ $x = -\frac{4}{3}$

54. $|2y - 7| < 3$

$-3 < 2y - 7 < 3$

$4 < 2y < 10$

$2 < y < 5$

56. $|1 + 2r| > 5$

$1 + 2r < -5$ or $1 + 2r > 5$

$2r < -6$ $2r > 4$

$r < -3$ $r > 2$

58. $|2t - 3| = |t - 5|$

$2t - 3 = t - 5$ or $2t - 3 = -(t - 5)$

$2t = t - 2$ $2t - 3 = -t + 5$

$t = -2$ $2t = -t + 8$

$3t = 8$

$t = \dfrac{8}{3}$

60. $|m + 3| = |m|$

$m + 3 = m$ or $m + 3 = -m$

$3 = 0$ $3 = -2m$

No solution $-\dfrac{3}{2} = m$

62. $|6 - q| \geq 1$

$6 - q \leq -1$ or $6 - q \geq 1$

$-q \leq -7$ $-q \geq -5$

$q \geq 7$ $q \leq 5$

64. $\left|\dfrac{2s + 1}{3}\right| = 5$

$\dfrac{2s + 1}{3} = 5$ or $\dfrac{2s + 1}{3} = -5$

$2s + 1 = 15$ $2s + 1 = -15$

$2s = 14$ $2s = -16$

$s = 7$ $s = -8$

66. $|1 - 4x| \leq 9$

$-9 \leq 1 - 4x \leq 9$

$-10 \leq -4x \leq 8$

$\dfrac{5}{2} \geq x \geq -2$

68. $|3z| = |z|$

$3z = z$ or $3z = -z$

$2z = 0$ $4z = 0$

$z = 0$ $z = 0$

70. $|p - 8| + 16 = 12$

$|p - 8| = -4$

No solution, since
absolute value is never
negative.

72. $|3 - w| = |w - 3|$

$3 - w = w - 3$ or $3 - w = -(w - 3)$

$-w = w - 6$ $3 - w = -w + 3$

$-2w = -6$ $0 = 0$

$w = 3$

Solution is all real numbers,
since $0 = 0$ is an identity.

74. $|r - 4| > 0$

If $r = 4$, then $|r - 4| = 0$. If $r \neq 4$, then $|r - 4| > 0$.
Therefore any real number except 4 is a solution.

76. $|r - 4| < 0$

No solution, since absolute value is never negative.

78. $|r - 4| \leq 0$

The only solution is 4.

80. a) $x = 1$ or $x = -1$ b) $-1 < x < 1$ c) $x < -1$ or $x > 1$

   $|x| = 1$ $|x| < 1$ $|x| > 1$

82. $|x - 3| = 8$ 84. $|x| \geq 5$

   $x - 3 = 8$ or $x - 3 = -8$ $x \leq -5$ or $x \geq 5$

   $x = 11$ $x = -5$

86. $|6 + 3x| < 9$

   $-9 < 6 + 3x < 9$

   $-15 < 3x < 3$

   $-5 < x < 1$

88. x = no. of records purchased

        $7x$ = nonmembers cost for x records

    $4x + 15$ = members cost for x records

    Difference in cost (in either order) < 12

                    $|7x - (4x + 15)| < 12$

                    $|7x - 4x - 15| < 12$

                        $|3x - 15| < 12$

                    $-12 < 3x - 15 < 12$

                        $3 < 3x < 27$

            1 record < x < 9 records

90. $\left|\dfrac{2.5x + 19.284}{0.815}\right| < 22.06$

$$-22.06 < \dfrac{2.5x + 19.284}{0.815} < 22.06$$

$$-17.9789 < 2.5x + 19.284 < 17.9789$$

$$-37.2629 < 2.5x < -1.3051$$

$$-14.90516 < x < -0.52204$$

NOTES

CHAPTER 4

RATIONAL EXPRESSIONS

Problem Set 4.1, pp. 145–146

2. $6x = 0$

$\quad x = 0$

4. $y - 2 = 0$

$\quad\quad y = 2$

6. $p + 3 = 0$

$\quad\quad p = -3$

8. Never undefined since denominator is never 0.

10. $\dfrac{7x}{8x} = \dfrac{7}{8}$

12. $\dfrac{7 + x}{8 + x}$ cannot be simplified.

14. $\dfrac{9b^2}{15} = \dfrac{3 \cdot 3b^2}{3 \cdot 5} = \dfrac{3b^2}{5}$

16. $\dfrac{-30c}{12c^2} = \dfrac{-6c \cdot 5}{6c \cdot 2c} = -\dfrac{5}{2c}$

18. $\dfrac{7xy^4}{14x^3y^2} = \dfrac{7xy^2 \cdot y^2}{7xy^2 \cdot 2x^2} = \dfrac{y^2}{2x^2}$

20. $\dfrac{x - 2}{7x - 14} = \dfrac{1(x - 2)}{7(x - 2)} = \dfrac{1}{7}$

22. $\dfrac{3py + pz}{3qy + qz} = \dfrac{p(3y + z)}{q(3y + z)} = \dfrac{p}{q}$

24. $\dfrac{r - s}{r^2 - s^2} = \dfrac{r - s}{(r + s)(r - s)} = \dfrac{1}{r + s}$

26. $\dfrac{12t + 20}{9t^2 - 25} = \dfrac{4(3t + 5)}{(3t - 5)(3t + 5)} = \dfrac{4}{3t - 5}$

28. $\dfrac{(w + 6)^3}{(w + 6)^6} = \dfrac{(w + 6)^3}{(w + 6)^3 (w + 6)^3} = \dfrac{1}{(w + 6)^3}$

30. $\dfrac{(x - y)^2}{x^2 - y^2} = \dfrac{(x - y)(x - y)}{(x + y)(x - y)} = \dfrac{x - y}{x + y}$

32. $\dfrac{10t^2 - 10t - 60}{6t^2 - 18t} = \dfrac{10(t^2 - t - 6)}{6t(t - 3)} = \dfrac{2 \cdot 5(t + 2)(t - 3)}{2 \cdot 3(t - 3)} = \dfrac{5(t + 2)}{3}$

34. $\dfrac{15p^2 + 2pq - 24q^2}{3p^2 - 5pq - 12q^2} = \dfrac{(3p + 4q)(5p - 6q)}{(3p + 4q)(p - 3q)} = \dfrac{5p - 6q}{p - 3q}$

36. $\dfrac{a + b}{b + a} = \dfrac{a + b}{a + b} = 1$

38. $\dfrac{a - b}{b - a} = \dfrac{a - b}{-(-b + a)} = \dfrac{1(a - b)}{-1(a - b)} = \dfrac{1}{-1} = -1$

40. $\dfrac{m - 1}{m^3 - 1} = \dfrac{m - 1}{(m - 1)(m^2 + m + 1)} = \dfrac{1}{m^2 + m + 1}$

42. $\dfrac{t - rst + s^2 - rs^3}{t - rst - s^2 + rs^3} = \dfrac{t(1 - rs) + s^2(1 - rs)}{t(1 - rs) - s^2(1 - rs)} = \dfrac{(1 - rs)(t + s^2)}{(1 - rs)(t - s^2)}$

$$= \dfrac{t + s^2}{t - s^2}$$

44. $\dfrac{c^3 - 64d^3}{4d^2 - cd - 8d + 2c} = \dfrac{c^3 - (4d)^3}{d(4d - c) - 2(4d - c)} = \dfrac{(c - 4d)(c^2 + 4cd + 16d^2)}{(4d - c)(d - 2)}$

$$= -\dfrac{c^2 + 4cd + 16d^2}{d - 2}$$

46. $\dfrac{(r - 3)2 - r}{(r - 3)r - 18} = \dfrac{2r - 6 - r}{r^2 - 3r - 18} = \dfrac{r - 6}{(r - 6)(r + 3)} = \dfrac{1}{r + 3}$

48. $\dfrac{(q^2 + 3)2 - 6}{(q^2 + 3)q} = \dfrac{2q^2 + 6 - 6}{(q^2 + 3)q} = \dfrac{2q \cdot q}{(q^2 + 3)q} = \dfrac{2q}{q^2 + 3}$

50. $\dfrac{3}{4x} = \dfrac{3 \cdot 9}{4x \cdot 9} = \dfrac{27}{36x}$

52. $\dfrac{1}{7x^3y^2} = \dfrac{1 \cdot 4y^7}{7x^3y^2 \cdot 4y^7} = \dfrac{4y^7}{28x^3y^9}$

54. $r = \dfrac{r}{1} = \dfrac{r(r - 8)}{1(r - 8)} = \dfrac{r^2 - 8r}{r - 8}$

56. $\dfrac{2m}{m - 2} = \dfrac{2m \cdot m}{(m - 2)m} = \dfrac{2m^2}{m(m - 2)}$

58. $\dfrac{s - 3}{s + 3} = \dfrac{(s - 3)4s}{(s + 3)4s} = \dfrac{4s^2 - 12s}{4s(s + 3)}$

60. $\dfrac{q + 1}{q - 6} = \dfrac{(q + 1)2(q + 3)}{(q - 6)2(q + 3)} = \dfrac{2q^2 + 8q + 6}{2(q - 6)(q + 3)}$

62. $p^2 - 64 = (p + 8)(p - 8)$

$$\frac{p - 5}{p + 8} = \frac{(p - 5)(p - 8)}{(p + 8)(p - 8)} = \frac{p^2 - 13p + 40}{(p + 8)(p - 8)}$$

64. $2x + 6 = 2(x + 3)$

$$\frac{17x^2}{x + 3} = \frac{17x^2 \cdot 2}{(x + 3)2} = \frac{34x^2}{2(x + 3)}$$

66. $\dfrac{3t - 1}{t + 5} = \dfrac{(3t - 1)(t + 5)}{(t + 5)(t + 5)} = \dfrac{3t^2 + 14t - 5}{(t + 5)^2}$

68. $w^3 - 8 = (w - 2)(w^2 + 2w + 4)$

$$\frac{w}{w - 2} = \frac{w(w^2 + 2w + 4)}{(w - 2)(w^2 + 2w + 4)} = \frac{w^3 + 2w^2 + 4w}{(w - 2)(w^2 + 2w + 4)}$$

70. $m = \dfrac{180(6) - 360}{6} = \dfrac{1080 - 360}{6} = \dfrac{720}{6} = 120°$

72. $y = \dfrac{x^3}{3x^2 - 3x + 1} = \dfrac{(0.57)^3}{3(0.57)^2 - 3(0.57) + 1} \approx 0.70$

Problem Set 4.2, pp. 149–151

2. $\dfrac{x^3}{15} \cdot \dfrac{3}{x} = \dfrac{3x^3}{15x} = \dfrac{x^2}{5}$

4. $\dfrac{8x^3}{20y} \cdot \dfrac{15y^2}{18x^5} = \dfrac{120x^3y^2}{360x^5y} = \dfrac{y}{3x^2}$

6. $\dfrac{4r}{5} \div \dfrac{3}{r} = \dfrac{4r}{5} \cdot \dfrac{r}{3} = \dfrac{4r^2}{15}$

8. $\dfrac{a^3b^2}{6b^3} \div \dfrac{a^2b^3}{6a^2} = \dfrac{a^3b^2}{6b^3} \cdot \dfrac{6a^2}{a^2b^3} = \dfrac{6a^5b^2}{6a^2b^6} = \dfrac{a^3}{b^4}$

10. $-\dfrac{4r}{15s^4} \cdot \dfrac{3r^3}{-14} \cdot \dfrac{-7s^2}{10r^4} = \dfrac{84r^4s^2}{-2100r^4s^4} = -\dfrac{1}{25s^2}$

12. $\dfrac{14ac^3}{-3(bc)^3} \div \dfrac{-7(ab)^2}{a^4b^2c^5} = \dfrac{14ac^3}{-3b^3c^3} \cdot \dfrac{a^4b^2c^5}{-7a^2b^2} = \dfrac{14a^5b^2c^8}{21a^2b^5c^3} = \dfrac{2a^3c^5}{3b^3}$

14. $\dfrac{11q}{3q-9} \cdot \dfrac{4q-12}{22} = \dfrac{11q}{3(q-3)} \cdot \dfrac{4(q-3)}{22} = \dfrac{2q}{3}$

16. $\dfrac{p^2-16}{p^3} \cdot \dfrac{8p^2+16p}{p^2-2p-8} = \dfrac{(p-4)(p+4)}{p^3} \cdot \dfrac{8p(p+2)}{(p-4)(p+2)} = \dfrac{8(p+4)}{p^2}$

18. $\dfrac{15n}{m} \div \dfrac{75}{n} = \dfrac{15n}{m} \cdot \dfrac{n}{75} = \dfrac{n^2}{5m}$

20. $\dfrac{xy}{xy-1} \div \dfrac{y}{x} = \dfrac{xy}{xy-1} \cdot \dfrac{x}{y} = \dfrac{x^2}{xy-1}$

22. $\dfrac{2r+6}{r^3+8r^2} \div \dfrac{3r+9}{r(2r+6)} = \dfrac{2(r+3)}{r^2(r+8)} \cdot \dfrac{2r(r+3)}{3(r+3)} = \dfrac{4(r+3)}{3r(r+8)}$

24. $\dfrac{s^2+3s-4}{s^2-6s+5} \cdot \dfrac{s-5}{7s^3+28s^2} = \dfrac{(s+4)(s-1)}{(s-5)(s-1)} \cdot \dfrac{s-5}{7s^2(s+4)} = \dfrac{1}{7s^2}$

26. $\dfrac{m^2+4m-12}{m^2+2m-8} \div \dfrac{m^2+7m+6}{m^2-1} = \dfrac{(m+6)(m-2)}{(m+4)(m-2)} \cdot \dfrac{(m+1)(m-1)}{(m+1)(m+6)}$

$$= \dfrac{m-1}{m+4}$$

28. $\dfrac{x+2}{x-2} \div (x^3+8) = \dfrac{x+2}{x-2} \cdot \dfrac{1}{x^3+8} = \dfrac{x+2}{x-2} \cdot \dfrac{1}{(x+2)(x^2-2x+4)}$

$$= \dfrac{1}{(x-2)(x^2-2x+4)}$$

30. $\dfrac{2c^2-9cd-5d^2}{c^2-d^2} \cdot \dfrac{(c+d)^2}{c^2-4cd-5d^2} = \dfrac{(2c+d)(c-5d)}{(c-d)(c+d)} \cdot \dfrac{(c+d)(c+d)}{(c-5d)(c+d)}$

$$= \dfrac{2c+d}{c-d}$$

32. $\dfrac{q + 3}{2} \div \dfrac{q^2 - 6q + 9}{5q + 15} = \dfrac{q + 3}{2} \cdot \dfrac{5(q + 3)}{(q - 3)^2} = \dfrac{5(q + 3)^2}{2(q - 3)^2}$

34. $\dfrac{20w^2 - 23w + 6}{8w^2 - 26w + 15} \cdot \dfrac{4w^2 - 25}{25w^2 - 4} = \dfrac{(5w - 2)(4w - 3)}{(2w - 5)(4w - 3)} \cdot \dfrac{(2w - 5)(2w + 5)}{(5w - 2)(5w + 2)}$

$$= \dfrac{2w + 5}{5w + 2}$$

36. $\dfrac{3u^2 + 7uv - 20v^2}{u^2 + 6uv + 8v^2} \div \dfrac{9u - 15v}{u^2v - 4v^3} = \dfrac{(3u - 5v)(u + 4v)}{(u + 2v)(u + 4v)} \cdot \dfrac{v(u - 2v)(u + 2v)}{3(3u - 5v)}$

$$= \dfrac{v(u - 2v)}{3}$$

38. $\dfrac{s - t}{s - 3t} \div \dfrac{s^2 + 3st + 9t^2}{s^3 - 27t^3} = \dfrac{s - t}{s - 3t} \cdot \dfrac{(s - 3t)(s^2 + 3st + 9t^2)}{s^2 + 3st + 9t^2}$

$$= s - t$$

40. $\dfrac{ms + mt - ns - nt}{m - n} \div (t^3 + s^3) = \dfrac{m(s + t) - n(s + t)}{m - n} \cdot \dfrac{1}{t^3 + s^3}$

$$= \dfrac{(s + t)(m - n)}{m - n} \cdot \dfrac{1}{(t + s)(t^2 - st + s^2)}$$

$$= \dfrac{1}{t^2 - st + s^2}$$

42. $\dfrac{ax + ay - bx - by}{b^5} \div \dfrac{b^2 - a^2}{b^2} = \dfrac{a(x + y) - b(x + y)}{b^5} \cdot \dfrac{b^2}{b^2 - a^2}$

$$= \dfrac{(x + y)(a - b)}{b^5} \cdot \dfrac{b^2}{(b - a)(b + a)}$$

$$= -\dfrac{x + y}{b^3(b + a)}$$

44. $\dfrac{1}{2u^2-6uv-5u+15v} \div \dfrac{1}{8u^3-125} = \dfrac{1}{2u\,(u-3v)-5\,(u-3v)} \cdot \dfrac{(2u)^3-5^3}{1}$

$= \dfrac{1}{(u-3v)\,(2u-5)} \cdot \dfrac{(2u-5)\,(4u^2+10u+25)}{1}$

$= \dfrac{4u^2+10u+25}{u-3v}$

46. $\left(\dfrac{w^4 - 2w^3}{4 - w^2} \div \dfrac{4 - w^2}{(w + 2)^2}\right) \cdot \dfrac{2z^3 - z^3 w}{2w^3 - w^3 z}$

$= \dfrac{w^3\,(w - 2)}{(2 + w)\,(2 - w)} \cdot \dfrac{(w + 2)\,(w + 2)}{(2 + w)\,(2 - w)} \cdot \dfrac{z^3\,(2 - w)}{w^3\,(2 - z)}$

$= -\,\dfrac{z^3}{2 - z}$

48. AC $= C \div x = \dfrac{0.5x^2 + 29x + 480}{x + 3} \div x$

$= \dfrac{0.5x^2 + 29x + 480}{x + 3} \cdot \dfrac{1}{x}$

$= \dfrac{0.5x^2 + 29x + 480}{x\,(x + 3)}$

50. $\dfrac{(2.37)^2 + 2.37 - 2}{(2.37)^2 - 4} \div \dfrac{1 - (2.37)^2}{(2.37)^2 - 2.37 - 2} = \dfrac{5.9869}{1.6169} \div \dfrac{-4.6169}{1.2469} = -1$

$\dfrac{m^2 + m - 2}{m^2 - 4} \div \dfrac{1 - m^2}{m^2 - m - 2} = \dfrac{(m + 2)\,(m - 1)}{(m + 2)\,(m - 2)} \cdot \dfrac{(m - 2)\,(m + 1)}{(1 - m)\,(1 + m)} = -1$

Problem Set 4.3, pp. 157-158

2. $\dfrac{9}{25r^2} + \dfrac{4}{25r^2} - \dfrac{3}{25r^2} = \dfrac{9 + 4 - 3}{25r^2} = \dfrac{10}{25r^2} = \dfrac{2}{5r^2}$

4. $\dfrac{k^2}{k^4} - \dfrac{k}{k^4} + \dfrac{7}{k^4} = \dfrac{k^2 - k + 7}{k^4}$

6. $\dfrac{p - 6}{p + 2} + \dfrac{8}{p + 2} = \dfrac{p - 6 + 8}{p + 2} = \dfrac{p + 2}{p + 2} = 1$

8. $\dfrac{7}{q} - \dfrac{q + 7}{q} = \dfrac{7 - (q + 7)}{q} = \dfrac{7 - q - 7}{q} = \dfrac{-q}{q} = -1$

10. $\dfrac{5x}{x^2 - 3x - 10} + \dfrac{10}{x^2 - 3x - 10} = \dfrac{5x + 10}{x^2 - 3x - 10} = \dfrac{5(x + 2)}{(x - 5)(x + 2)} = \dfrac{5}{x - 5}$

12. $\dfrac{m^2 + 25n^2}{m - 5n} - \dfrac{10mn}{m - 5n} = \dfrac{m^2 + 25n^2 - 10mn}{m - 5n} = \dfrac{m^2 - 10mn + 25n^2}{m - 5n}$

$$= \dfrac{(m - 5n)^2}{m - 5n}$$

$$= m - 5n$$

14. $\dfrac{x(x + 4)}{(x - 4)(x + 4)} - \dfrac{4(x - 4)}{(x + 4)(x - 4)} = \dfrac{x(x + 4) - 4(x - 4)}{(x - 4)(x + 4)} = \dfrac{x^2 + 4x - 4x + 16}{(x - 4)(x + 4)}$

$$= \dfrac{x^2 + 16}{(x - 4)(x + 4)}$$

16. $\dfrac{s^2 - 64}{s - 8} - \dfrac{64}{8 - s} = \dfrac{s^2 - 64}{s - 8} + \dfrac{64}{s - 8} = \dfrac{s^2 - 64 + 64}{s - 8} = \dfrac{s^2}{s - 8}$

18. $\dfrac{c}{c^2 - d^2} + \dfrac{d}{d^2 - c^2} = \dfrac{c}{c^2 - d^2} + \dfrac{-d}{c^2 - d^2} = \dfrac{c + (-d)}{c^2 - d^2} = \dfrac{c - d}{(c + d)(c - d)}$

$$= \dfrac{1}{c + d}$$

20. a) LCD = 14a

    b) $\dfrac{1}{2a} + \dfrac{1}{7a} = \dfrac{1 \cdot 7}{2a \cdot 7} + \dfrac{1 \cdot 2}{7a \cdot 2} = \dfrac{7}{14a} + \dfrac{2}{14a} = \dfrac{9}{14a}$

22. a) LCD = b^3

    b) $\dfrac{3}{b} - \dfrac{1}{b^3} = \dfrac{3 \cdot b^2}{b \cdot b^2} - \dfrac{1}{b^3} = \dfrac{3b^2}{b^3} - \dfrac{1}{b^3} = \dfrac{3b^2 - 1}{b^3}$

24. a)　$LCD = 16c^2$

b)　$\dfrac{5}{16c} + \dfrac{3}{4c^2} = \dfrac{5 \cdot c}{16c \cdot c} + \dfrac{3 \cdot 4}{4c^2 \cdot 4} = \dfrac{5c}{16c^2} + \dfrac{12}{16c^2}$

$$= \dfrac{5c + 12}{16c^2}$$

26. a)　$27x^2y^2 = 3^3x^2y^2$

$60xy^4 = 2^2 \cdot 3 \cdot 5xy^4$

$LCD = 2^2 \cdot 3^3 \cdot 5x^2y^4 = 540x^2y^4$

b)　$\dfrac{1}{27x^2y^2} + \dfrac{7}{60xy^4} = \dfrac{1 \cdot 20y^2}{27x^2y^2 \cdot 20y^2} + \dfrac{7 \cdot 9x}{60xy^4 \cdot 9x}$

$$= \dfrac{20y^2}{540x^2y^4} + \dfrac{63x}{540x^2y^4}$$

$$= \dfrac{20y^2 + 63x}{540x^2y^4}$$

28. a)　$LCD = k$

b)　$\dfrac{1}{k} - k = \dfrac{1}{k} - \dfrac{k \cdot k}{1 \cdot k} = \dfrac{1}{k} - \dfrac{k^2}{k} = \dfrac{1 - k^2}{k}$

30. a)　$LCD = p + 2$

b)　$p - \dfrac{p^2}{p + 2} = \dfrac{p(p + 2)}{1(p + 2)} - \dfrac{p^2}{p + 2} = \dfrac{p^2 + 2p}{p + 2} - \dfrac{p^2}{p + 2}$

$$= \dfrac{p^2 + 2p - p^2}{p + 2}$$

$$= \dfrac{2p}{p + 2}$$

32. a) LCD = $q(q + 3)$

b) $\dfrac{18}{q(q + 3)} + \dfrac{6}{q + 3} = \dfrac{18}{q(q + 3)} + \dfrac{6q}{(q + 3)q} = \dfrac{6q + 18}{q(q + 3)}$

$= \dfrac{6(q + 3)}{q(q + 3)}$

$= \dfrac{6}{q}$

34. a) LCD = $(h - 12)(h - 4)$

b) $\dfrac{12}{h - 12} - \dfrac{4}{h - 4} = \dfrac{12(h - 4)}{(h - 12)(h - 4)} - \dfrac{4(h - 12)}{(h - 4)(h - 12)}$

$= \dfrac{12(h - 4) - 4(h - 12)}{(h - 12)(h - 4)}$

$= \dfrac{12h - 48 - 4h + 48}{(h - 12)(h - 4)}$

$= \dfrac{8h}{(h - 12)(h - 4)}$

36. a) LCD = $(r - 2)(r + 1)$

b) $\dfrac{r}{r - 2} - \dfrac{r - 5}{r + 1} = \dfrac{r(r + 1)}{(r - 2)(r + 1)} - \dfrac{(r - 5)(r - 2)}{(r + 1)(r - 2)}$

$= \dfrac{r(r + 1) - (r - 5)(r - 2)}{(r - 2)(r + 1)}$

$= \dfrac{r^2 + r - (r^2 - 7r + 10)}{(r - 2)(r + 1)}$

$= \dfrac{r^2 + r - r^2 + 7r - 10}{(r - 2)(r + 1)}$

$= \dfrac{8r - 10}{(r - 2)(r + 1)}$

$= \dfrac{2(4r - 5)}{(r - 2)(r + 1)}$

38. a) $s + 6 = s + 6$

   $3s + 18 = 3(s + 6)$

   $LCD = 3(s + 6)$

   b) $\dfrac{-4}{s + 6} - \dfrac{2s}{3s + 18} = \dfrac{-4 \cdot 3}{(s + 6)3} - \dfrac{2s}{3(s + 6)} = \dfrac{-12 - 2s}{3(s + 6)}$

   $= \dfrac{-2(s + 6)}{3(s + 6)}$

   $= -\dfrac{2}{3}$

40. a) $x^2 - 49 = (x - 7)(x + 7)$

   $x + 7 = x + 7$

   $LCD = (x - 7)(x + 7)$

   b) $\dfrac{20}{x^2 - 49} + \dfrac{3}{x + 7} = \dfrac{20}{(x - 7)(x + 7)} + \dfrac{3(x - 7)}{(x + 7)(x - 7)} = \dfrac{20 + 3(x - 7)}{(x + 7)(x - 7)}$

   $= \dfrac{20 + 3x - 21}{(x + 7)(x - 7)}$

   $= \dfrac{3x - 1}{(x + 7)(x - 7)}$

42. a) $2y + 16 = 2(y + 8)$

   $y^2 - 64 = (y + 8)(y - 8)$

   $LCD = 2(y + 8)(y - 8)$

   b) $\dfrac{1}{2y + 16} - \dfrac{y}{y^2 - 64} = \dfrac{1}{2(y + 8)} - \dfrac{y}{(y + 8)(y - 8)}$

   $= \dfrac{1(y - 8)}{2(y + 8)(y - 8)} - \dfrac{y \cdot 2}{(y + 8)(y - 8)2}$

   $= \dfrac{y - 8 - 2y}{2(y + 8)(y - 8)}$

   $= \dfrac{-y - 8}{2(y + 8)(y - 8)}$

   $= \dfrac{-(y + 8)}{2(y + 8)(y - 8)}$

   $= \dfrac{-1}{2(y - 8)}$

44. a) $\quad m^2 + m = m(m + 1)$

$m^2 + 2m + 1 = (m + 1)^2$

$\text{LCD} = m(m + 1)^2$

b) $\dfrac{5}{m^2 + m} - \dfrac{5}{m^2 + 2m + 1} = \dfrac{5}{m(m + 1)} - \dfrac{5}{(m + 1)^2}$

$= \dfrac{5(m + 1)}{m(m + 1)^2} - \dfrac{5m}{(m + 1)^2 m}$

$= \dfrac{5(m + 1) - 5m}{m(m + 1)^2}$

$= \dfrac{5m + 5 - 5m}{m(m + 1)^2}$

$= \dfrac{5}{m(m + 1)^2}$

46. a) $\quad ab - b^2 = b(a - b)$

$a^2 - b^2 = (a - b)(a + b)$

$\text{LCD} = b(a - b)(a + b)$

b) $\dfrac{a}{ab - b^2} - \dfrac{b^2}{a^2 - b^2} = \dfrac{a}{b(a - b)} - \dfrac{b^2}{(a - b)(a + b)}$

$= \dfrac{a(a + b)}{b(a - b)(a + b)} - \dfrac{b^2 \cdot b}{(a - b)(a + b)b}$

$= \dfrac{a(a + b) - b^3}{b(a - b)(a + b)}$

$= \dfrac{a^2 + ab - b^3}{b(a - b)(a + b)}$

48.   a) $k^2 + 3k - 10 = (k + 5)(k - 2)$

$k^2 - k - 2 = (k + 1)(k - 2)$

LCD $= (k + 5)(k - 2)(k + 1)$

b) $\dfrac{1}{k^2 + 3k - 10} + \dfrac{1}{k^2 - k - 2} = \dfrac{1}{(k + 5)(k - 2)} + \dfrac{1}{(k + 1)(k - 2)}$

$= \dfrac{1(k+1)}{(k+5)(k-2)(k+1)} + \dfrac{1(k+5)}{(k+1)(k-2)(k+5)}$

$= \dfrac{(k + 1) + (k + 5)}{(k + 5)(k - 2)(k + 1)}$

$= \dfrac{2k + 6}{(k + 5)(k - 2)(k + 1)}$

$= \dfrac{2(k + 3)}{(k + 5)(k - 2)(k + 1)}$

50.   a) $5x + 10 = 5(x + 2)$

$x^2 - 4 = (x + 2)(x - 2)$

$x - 2 = x - 2$

LCD $= 5(x + 2)(x - 2)$

b) $\dfrac{x}{5x + 10} - \dfrac{x + 6}{x^2 - 4} + \dfrac{2}{x - 2}$

$= \dfrac{x}{5(x + 2)} - \dfrac{x + 6}{(x + 2)(x - 2)} + \dfrac{2}{x - 2}$

$= \dfrac{x(x - 2)}{5(x + 2)(x - 2)} - \dfrac{(x + 6)5}{(x + 2)(x - 2)5} + \dfrac{2(x + 2)5}{(x - 2)(x + 2)5}$

$= \dfrac{x(x - 2) - (x + 6)5 + 2(x + 2)5}{5(x + 2)(x - 2)}$

$= \dfrac{x^2 - 2x - 5x - 30 + 10x + 20}{5(x + 2)(x - 2)}$

$= \dfrac{x^2 + 3x - 10}{5(x + 2)(x - 2)}$

$= \dfrac{(x + 5)(x - 2)}{5(x + 2)(x - 2)}$

$= \dfrac{x + 5}{5(x + 2)}$

52. a) $t + 3 = t + 3$

              $t - 4 = t - 4$

         $t^2 - t - 12 = (t + 3)(t - 4)$

              LCD $= (t + 3)(t - 4)$

  b) $\dfrac{t - 2}{t + 3} + \dfrac{t - 5}{t - 4} - \dfrac{t^2 - 23}{t^2 - t - 12}$

  $= \dfrac{(t - 2)(t - 4)}{(t + 3)(t - 4)} + \dfrac{(t - 5)(t + 3)}{(t - 4)(t + 3)} - \dfrac{t^2 - 23}{(t + 3)(t - 4)}$

  $= \dfrac{(t - 2)(t - 4) + (t - 5)(t + 3) - (t^2 - 23)}{(t + 3)(t - 4)}$

  $= \dfrac{t^2 - 6t + 8 + t^2 - 2t - 15 - t^2 + 23}{(t + 3)(t - 4)}$

  $= \dfrac{t^2 - 8t + 16}{(t + 3)(t - 4)}$

  $= \dfrac{(t - 4)(t - 4)}{(t + 3)(t - 4)}$

  $= \dfrac{t - 4}{t + 3}$

54. a) LCD $= (a - 1)^2$

  b) $1 + \dfrac{1}{a - 1} + \dfrac{a}{(a - 1)^2} = \dfrac{1(a - 1)^2}{(a - 1)^2} + \dfrac{1(a - 1)}{(a - 1)(a - 1)} + \dfrac{a}{(a - 1)^2}$

  $= \dfrac{(a - 1)^2 + (a - 1) + a}{(a - 1)^2}$

  $= \dfrac{a^2 - 2a + 1 + a - 1 + a}{(a - 1)^2}$

  $= \dfrac{a^2}{(a - 1)^2}$

56. a) LCD = r + 1

   b) $\dfrac{5r}{r + 1} - r - 3 = \dfrac{5r}{r + 1} - \dfrac{r + 3}{1} = \dfrac{5r}{r + 1} - \dfrac{(r + 3)(r + 1)}{1(r + 1)}$

   $= \dfrac{5r - (r + 3)(r + 1)}{r + 1}$

   $= \dfrac{5r - (r^2 + 4r + 3)}{r + 1}$

   $= \dfrac{5r - r^2 - 4r - 3}{r + 1}$

   $= \dfrac{-r^2 + r - 3}{r + 1}$

58. a) $(s - 6)^2 = (s - 6)^2$

   $s^2 - 36 = (s - 6)(s + 6)$

   LCD $= (s - 6)^2(s + 6)$

   b) $\dfrac{s}{(s - 6)^2} + \dfrac{s}{s^2 - 36} = \dfrac{s}{(s - 6)^2} + \dfrac{s}{(s - 6)(s + 6)}$

   $= \dfrac{s(s + 6)}{(s - 6)^2(s + 6)} + \dfrac{s(s - 6)}{(s - 6)(s + 6)(s - 6)}$

   $= \dfrac{s(s + 6) + s(s - 6)}{(s - 6)^2(s + 6)}$

   $= \dfrac{s^2 + 6s + s^2 - 6s}{(s - 6)^2(s + 6)}$

   $= \dfrac{2s^2}{(s - 6)^2(s + 6)}$

60. a) $x - y = x - y$

   $x^3 - y^3 = (x - y)(x^2 + xy + y^2)$

   LCD $= (x - y)(x^2 + xy + y^2)$

   b) $\dfrac{1}{x - y} - \dfrac{xy}{x^3 - y^3} = \dfrac{1}{x - y} - \dfrac{xy}{(x - y)(x^2 + xy + y^2)}$

   $= \dfrac{1(x^2 + xy + y^2)}{(x - y)(x^2 + xy + y^2)} - \dfrac{xy}{(x - y)(x^2 + xy + y^2)}$

$$= \frac{(x^2 + xy + y^2) - xy}{(x - y)(x^2 + xy + y^2)}$$

$$= \frac{x^2 + y^2}{(x - y)(x^2 + xy + y^2)}$$

62. a) $LCD = (c + 3d)(c - 3d)$

b) $\left(\dfrac{d}{c + 3d} + \dfrac{d}{c - 3d}\right) \div \dfrac{cd}{6c^2 - 13cd - 15d^2}$

$$= \left(\frac{d(c - 3d)}{(c + 3d)(c - 3d)} + \frac{d(c + 3d)}{(c - 3d)(c + 3d)}\right) \cdot \frac{6c^2 - 13cd - 15d^2}{cd}$$

$$= \frac{cd - 3d^2 + cd + 3d^2}{(c + 3d)(c - 3d)} \cdot \frac{(6c + 5d)(c - 3d)}{cd}$$

$$= \frac{2cd}{(c + 3d)(c - 3d)} \cdot \frac{(6c + 5d)(c - 3d)}{cd}$$

$$= \frac{2(6c + 5d)}{c + 3d}$$

64. a) $m^2 - m - 2 = (m - 2)(m + 1)$

$m^2 - 4m + 4 = (m - 2)^2$

$m^2 + m = m(m + 1)$

$LCD = m(m - 2)^2(m + 1)$

b) $\dfrac{m}{m^2 - m - 2} - \dfrac{m + 2}{m^2 - 4m + 4} - \dfrac{1}{m^2 + m}$

$$= \frac{m}{(m - 2)(m + 1)} - \frac{m + 2}{(m - 2)^2} - \frac{1}{m(m + 1)}$$

$$= \frac{m(m - 2)m}{(m - 2)(m + 1)(m - 2)m} - \frac{(m + 2)(m + 1)m}{(m - 2)^2(m + 1)m} - \frac{1(m - 2)^2}{m(m + 1)(m - 2)^2}$$

$$= \frac{m^2(m - 2) - (m^2 + 3m + 2)m - (m^2 - 4m + 4)}{m(m - 2)^2(m + 1)}$$

$$= \frac{m^3 - 2m^2 - m^3 - 3m^2 - 2m - m^2 + 4m - 4}{m(m - 2)^2(m + 1)}$$

$$= \frac{-6m^2 + 2m - 4}{m(m - 2)^2(m + 1)}$$

$$= \frac{-2(3m^2 - m + 2)}{m(m - 2)^2(m + 1)}$$

66. $\dfrac{1}{n} + \dfrac{1}{n+2} = \dfrac{1(n+2)}{n(n+2)} + \dfrac{1 \cdot n}{(n+2)n} = \dfrac{(n+2)+n}{n(n+2)} = \dfrac{2n+2}{n(n+2)} = \dfrac{2(n+}{n(n+}$

68. $v^2 = Mg\left(\dfrac{2}{R} + \dfrac{1}{a}\right)$

$v^2 = Mg\left(\dfrac{2 \cdot a}{R \cdot a} + \dfrac{1 \cdot R}{a \cdot R}\right)$

$v^2 = \dfrac{Mg}{1} \cdot \dfrac{2a+R}{aR}$

$v^2 = \dfrac{Mg(2a+R)}{aR}$

70. $\left(1 - \dfrac{1}{3.75}\right)\left(1 - \dfrac{1}{3.75-1}\right)\left(1 - \dfrac{1}{3.75-2}\right) = (0.7\overline{3})(0.6\overline{3})(0.\overline{428571}) =$

$\left(1 - \dfrac{1}{n}\right)\left(1 - \dfrac{1}{n-1}\right)\left(1 - \dfrac{1}{n-2}\right) = \left(\dfrac{n}{n} - \dfrac{1}{n}\right)\left(\dfrac{n-1}{n-1} - \dfrac{1}{n-1}\right)\left(\dfrac{n-2}{n-2} - \dfrac{1}{n-}\right)$

$= \dfrac{n-1}{n} \cdot \dfrac{n-1-1}{n-1} \cdot \dfrac{n-2-1}{n-2}$

$= \dfrac{n-1}{n} \cdot \dfrac{n-2}{n-1} \cdot \dfrac{n-3}{n-2}$

$= \dfrac{n-3}{n}$

$\dfrac{n-3}{n} = \dfrac{3.75-3}{3.75} = \dfrac{0.75}{3.75} = 0.2$

Problem Set 4.4, pp. 162-163

2. a) $\dfrac{u}{\frac{v}{w}} = u \div \dfrac{v}{w} = \dfrac{u}{1} \cdot \dfrac{w}{v} = \dfrac{uw}{v}$

   b) $\dfrac{\frac{u}{v}}{w} = \dfrac{u}{v} \div w = \dfrac{u}{v} \cdot \dfrac{1}{w} = \dfrac{u}{vw}$

4. $\dfrac{\frac{5}{8} + \frac{2}{3}}{\frac{5}{6} + \frac{1}{2}} = \dfrac{\frac{15}{24} + \frac{16}{24}}{\frac{5}{6} + \frac{3}{6}} = \dfrac{\frac{31}{24}}{\frac{8}{6}} = \dfrac{31}{24} \cdot \dfrac{6}{8} = \dfrac{31}{32}$

6. $\dfrac{2d}{\frac{d}{15} + \frac{d}{6}} = \dfrac{2d}{\frac{2d}{30} + \frac{5d}{30}} = \dfrac{2d}{\frac{7d}{30}} = \dfrac{2d}{1} \cdot \dfrac{30}{7d} = \dfrac{60}{7}$

8. $\dfrac{\dfrac{x+2}{x}}{\dfrac{x-2}{5x}} = \dfrac{x+2}{x} \cdot \dfrac{5x}{x-2} = \dfrac{5(x+2)}{x-2}$

10. $\dfrac{\dfrac{p^3}{p^2-9}}{\dfrac{p^2}{p-3}} = \dfrac{p^3}{p^2-9} \cdot \dfrac{p-3}{p^2} = \dfrac{p^3}{(p+3)(p-3)} \cdot \dfrac{p-3}{p^2} = \dfrac{p}{p+3}$

12. $\dfrac{\dfrac{2}{r}+\dfrac{2}{s}}{\dfrac{2}{r}-\dfrac{2}{s}} = \dfrac{(\dfrac{2}{r}+\dfrac{2}{s})rs}{(\dfrac{2}{r}-\dfrac{2}{s})rs} = \dfrac{2s+2r}{2s-2r} = \dfrac{2(s+r)}{2(s-r)} = \dfrac{s+r}{s-r}$

14. $\dfrac{1+\dfrac{1}{xy}}{\dfrac{x^2}{y^2}+\dfrac{x}{y^3}} = \dfrac{(1+\dfrac{1}{xy})xy^3}{(\dfrac{x^2}{y^2}+\dfrac{x}{y^3})xy^3} = \dfrac{xy^3+y^2}{x^3y+x^2} = \dfrac{y^2(xy+1)}{x^2(xy+1)} = \dfrac{y^2}{x^2}$

16. $\dfrac{25-\dfrac{1}{m^2}}{5-\dfrac{1}{m}} = \dfrac{(25-\dfrac{1}{m^2})m^2}{(5-\dfrac{1}{m})m^2} = \dfrac{25m^2-1}{5m^2-m} = \dfrac{(5m+1)(5m-1)}{m(5m-1)} = \dfrac{5m+1}{m}$

18. $\dfrac{1-\dfrac{a^2}{b^2}}{\dfrac{a}{8b^2}-\dfrac{1}{8b}} = \dfrac{(1-\dfrac{a^2}{b^2})8b^2}{(\dfrac{a}{8b^2}-\dfrac{1}{8b})8b^2} = \dfrac{8b^2-8a^2}{a-b} = \dfrac{8(b-a)(b+a)}{a-b} = -8(b+a)$

20. $\dfrac{\dfrac{c}{4}-\dfrac{1}{c}}{1+\dfrac{c+4}{c}} = \dfrac{(\dfrac{c}{4}-\dfrac{1}{c})4c}{(1+\dfrac{c+4}{c})4c} = \dfrac{c^2-4}{4c+(c+4)4} = \dfrac{(c+2)(c-2)}{8c+16}$

$= \dfrac{(c+2)(c-2)}{8(c+2)} = \dfrac{c-2}{8}$

22. $\dfrac{15-\dfrac{17}{p}-\dfrac{4}{p^2}}{3+\dfrac{5}{p}-\dfrac{12}{p^2}} = \dfrac{(15-\dfrac{17}{p}-\dfrac{4}{p^2})p^2}{(3+\dfrac{5}{p}-\dfrac{12}{p^2})p^2} = \dfrac{15p^2-17p-4}{3p^2+5p-12}$

$= \dfrac{(3p-4)(5p+1)}{(3p-4)(p+3)}$

$= \dfrac{5p+1}{p+3}$

24. $\dfrac{1 - \dfrac{7}{d+1}}{\dfrac{4}{1+d} + 1} = \dfrac{(1 - \dfrac{7}{d+1})(d+1)}{(\dfrac{4}{1+d} + 1)(d+1)} = \dfrac{1(d+1) - 7}{4 + 1(d+1)} = \dfrac{d-6}{d+5}$

26. $\dfrac{x-y}{\dfrac{1}{y} - \dfrac{1}{x}} = \dfrac{(x-y)xy}{(\dfrac{1}{y} - \dfrac{1}{x})xy} = \dfrac{(x-y)xy}{x-y} = xy$

28. $\dfrac{1 + \dfrac{2}{t}}{t^3 + 8} = \dfrac{(1 + \dfrac{2}{t})t}{(t^3 + 8)t} = \dfrac{t+2}{(t+2)(t^2 - 2t + 4)t} = \dfrac{1}{t(t^2 - 2t + 4)}$

30. $\dfrac{\dfrac{5}{m} + \dfrac{1}{n}}{\dfrac{1}{n^3} + \dfrac{125}{m^3}} = \dfrac{(\dfrac{5}{m} + \dfrac{1}{n})m^3 n^3}{(\dfrac{1}{n^3} + \dfrac{125}{m^3})m^3 n^3} = \dfrac{5m^2 n^3 + m^3 n^2}{m^3 + 125 n^3}$

$= \dfrac{m^2 n^2 (5n + m)}{(m + 5n)(m^2 - 5mn + 25n^2)}$

$= \dfrac{m^2 n^2}{m^2 - 5mn + 25n^2}$

32. $\dfrac{\dfrac{1}{u^2} - \dfrac{1}{v^2}}{\dfrac{1}{u^3} + \dfrac{1}{v^3}} = \dfrac{(\dfrac{1}{u^2} - \dfrac{1}{v^2})u^3 v^3}{(\dfrac{1}{u^3} + \dfrac{1}{v^3})u^3 v^3} = \dfrac{uv^3 - u^3 v}{v^3 + u^3}$

$= \dfrac{uv(v - u)(v + u)}{(v + u)(v^2 - vu + u^2)}$

$= \dfrac{uv(v - u)}{v^2 - vu + u^2}$

34. $\dfrac{1 - \dfrac{1}{q}}{1 - \dfrac{2q-1}{q^2}} = \dfrac{(1 - \dfrac{1}{q})q^2}{(1 - \dfrac{2q-1}{q^2})q^2} = \dfrac{q^2 - q}{q^2 - (2q-1)} = \dfrac{q(q-1)}{q^2 - 2q + 1}$

$= \dfrac{q(q-1)}{(q-1)^2}$

$= \dfrac{q}{q-1}$

36.

$$\frac{k + 5 + \dfrac{7}{k - 3}}{k + 3 - \dfrac{7}{k - 3}} = \frac{(k + 5 + \dfrac{7}{k - 3})(k - 3)}{(k + 3 - \dfrac{7}{k - 3})(k - 3)} = \frac{(k + 5)(k - 3) + 7}{(k + 3)(k - 3) - 7}$$

$$= \frac{k^2 + 2k - 15 + 7}{k^2 - 9 - 7}$$

$$= \frac{k^2 + 2k - 8}{k^2 - 16}$$

$$= \frac{(k - 2)(k + 4)}{(k - 4)(k + 4)}$$

$$= \frac{k - 2}{k - 4}$$

38.

$$\frac{\dfrac{w - 3}{w + 3} + \dfrac{w + 3}{w - 3}}{\dfrac{w - 3}{w + 3} - \dfrac{w + 3}{w - 3}} = \frac{(\dfrac{w - 3}{w + 3} + \dfrac{w + 3}{w - 3})(w + 3)(w - 3)}{(\dfrac{w - 3}{w + 3} - \dfrac{w + 3}{w - 3})(w + 3)(w - 3)}$$

$$= \frac{(w - 3)(w - 3) + (w + 3)(w + 3)}{(w - 3)(w - 3) - (w + 3)(w + 3)}$$

$$= \frac{w^2 - 6w + 9 + w^2 + 6w + 9}{w^2 - 6w + 9 - (w^2 + 6w + 9)}$$

$$= \frac{2w^2 + 18}{-12w}$$

$$= -\frac{2(w^2 + 9)}{12w}$$

$$= -\frac{w^2 + 9}{6w}$$

40.

$$x + \frac{1}{1 + \dfrac{1}{x+1}} = x + \frac{1}{\dfrac{x + 1}{x + 1} + \dfrac{1}{x + 1}} = x + \frac{1}{\dfrac{x + 2}{x + 1}}$$

$$= x + \frac{x + 1}{x + 2}$$

$$= \frac{x(x + 2)}{x + 2} + \frac{x + 1}{x + 2}$$

$$= \frac{x^2 + 2x + x + 1}{x + 2}$$

$$= \frac{x^2 + 3x + 1}{x + 2}$$

42. $$2 - \cfrac{1}{2 - \cfrac{1}{2 - \cfrac{1}{2-1}}} = 2 - \cfrac{1}{2 - \cfrac{1}{2 - \frac{1}{1}}} = 2 - \cfrac{1}{2 - \frac{1}{1}}$$

$$= 2 - \frac{1}{1}$$

$$= 1$$

44. Let d = distance to work.

$$\text{Average rate} = \frac{\text{total distance}}{\text{total time}} = \frac{2d}{\dfrac{d}{30} + \dfrac{d}{50}}$$

$$= \frac{2d}{\dfrac{5d}{150} + \dfrac{3d}{150}}$$

$$= \frac{2d}{\dfrac{8d}{150}}$$

$$= \frac{2d}{1} \cdot \frac{150}{8d}$$

$$= 37.5 \text{ mph}$$

46. Let d = distance up the slope.

$$\text{Average rate} = \frac{\text{total distance}}{\text{total time}} = \frac{2d}{\dfrac{d}{2} + \dfrac{d}{35}}$$

$$= \frac{2d}{\dfrac{35d}{70} + \dfrac{2d}{70}}$$

$$= \frac{2d}{\dfrac{37d}{70}}$$

$$= \frac{2d}{1} \cdot \frac{70}{37d}$$

$$= \frac{140}{37}$$

$$\approx 3.78 \text{ mph}$$

Problem Set 4.5, pp. 168-170

2. $\frac{x}{4} + \frac{x}{5} = 18$

$20(\frac{x}{4} + \frac{x}{5}) = 20(18)$

$5x + 4x = 360$

$9x = 360$

$x = 4$

4. $\frac{x}{5} + \frac{x}{2} = -7$

$10(\frac{x}{5} + \frac{x}{2}) = 10(-7)$

$2x + 5x = -70$

$7x = -70$

$x = -10$

6. $\frac{1}{4x} - \frac{1}{3x} = \frac{1}{6}$

$12x(\frac{1}{4x} - \frac{1}{3x}) = 12x(\frac{1}{6})$

$3 - 4 = 2x$

$-1 = 2x$

$-\frac{1}{2} = x$

8. $\frac{2}{y} = \frac{1}{3} - \frac{1}{5}$

$15y(\frac{2}{y}) = 15y(\frac{1}{3} - \frac{1}{5})$

$30 = 5y - 3y$

$30 = 2y$

$15 = y$

10. $2 - \frac{4}{3k} = \frac{2}{k} - \frac{4}{3}$

$3k(2 - \frac{4}{3k}) = 3k(\frac{2}{k} - \frac{4}{3})$

$6k - 4 = 6 - 4k$

$6k = 10 - 4k$

$10k = 10$

$k = 1$

12. $\frac{7x - 2}{x + 6} = 0$

$(x + 6)\frac{7x - 2}{x + 6} = (x + 6) \cdot 0$

$7x - 2 = 0$

$7x = 2$

$x = \frac{2}{7}$

14. $\frac{3t + 19}{t - 3} = 1$

$(t - 3)\frac{3t + 19}{t - 3} = (t - 3)1$

$3t + 19 = t - 3$

$3t = t - 22$

$2t = -22$

$t = -11$

16. $\frac{5}{u + 2} = \frac{3}{u - 2}$

$5(u - 2) = (u + 2)3$

$5u - 10 = 3u + 6$

$5u = 3u + 16$

$2u = 16$

$u = 8$

18. $\dfrac{3}{5v - 2} = \dfrac{5}{3v + 4}$

$3(3v + 4) = (5v - 2)5$

$9v + 12 = 25v - 10$

$9v = 25v - 22$

$-16v = -22$

$v = \dfrac{-22}{-16}$

$v = \dfrac{11}{8}$

20. $\dfrac{w}{w - 4} = \dfrac{w}{w + 4}$

$w(w + 4) = (w - 4)w$

$w^2 + 4w = w^2 - 4w$

$4w = -4w$

$8w = 0$

$w = 0$

22. $\dfrac{2s}{s + 8} = \dfrac{10}{s + 8}$

$(s + 8)\dfrac{2s}{s + 8} = (s + 8)\dfrac{10}{s + 8}$

$2s = 10$

$s = 5$

24. $\dfrac{x}{x - 7} + 1 = \dfrac{x + 11}{x - 7}$

$(x - 7)\dfrac{x}{x - 7} + (x - 7)1 = (x - 7)\dfrac{x + 11}{x - 7}$

$x + x - 7 = x + 11$

$2x - 7 = x + 11$

$2x = x + 18$

$x = 18$

26. $\dfrac{6}{r} + 3 = \dfrac{3r}{r - 3}$

$r(r - 3)\dfrac{6}{r} + r(r - 3)3 = r(r - 3)\dfrac{3r}{r - 3}$

$(r - 3)6 + 3r(r - 3) = 3r^2$

$6r - 18 + 3r^2 - 9r = 3r^2$

$3r^2 - 3r - 18 = 3r^2$

$-3r - 18 = 0$

$-3r = 18$

$r = -6$

28.
$$\frac{3}{s} + \frac{2}{s-1} = \frac{4}{s} \qquad LCD = s(s-1)$$

$$s(s-1)\frac{3}{s} + s(s-1)\frac{2}{s-1} = s(s-1)\frac{4}{s}$$

$$(s-1)3 + 2s = (s-1)4$$

$$3s - 3 + 2s = 4s - 4$$

$$-3 + 5s = 4s - 4$$

$$5s = 4s - 1$$

$$s = -1$$

30.
$$\frac{t+2}{2t+8} + \frac{t}{t+4} = \frac{1}{2}$$

$$2(t+4)\frac{t+2}{2(t+4)} + 2(t+4)\frac{t}{t+4} = 2(t+4)\frac{1}{2}$$

$$t + 2 + 2t = t + 4$$

$$2 + 3t = t + 4$$

$$3t = t + 2$$

$$2t = 2$$

$$t = 1$$

32.
$$\frac{x}{x-3} + 3 = \frac{x}{x-3}$$

$$(x-3)\frac{x}{x-3} + (x-3)3 = (x-3)\frac{x}{x-3}$$

$$x + 3x - 9 = x$$

$$4x - 9 = x$$

$$-9 = -3x$$

$$3 = x$$

But 3 does not check because it makes both denominators in the
original equation 0. Therefore, there is no solution.

34.
$$\frac{4}{y+2} - \frac{5}{y-2} = \frac{y}{y^2-4}$$

$$(y+2)(y-2)\frac{4}{y+2} - (y+2)(y-2)\frac{5}{y-2} = (y+2)(y-2)\frac{y}{(y+2)(y-}$$

$$(y-2)4 - (y+2)5 = y$$

$$4y - 8 - 5y - 10 = y$$

$$-y - 18 = y$$

$$-18 = 2y$$

$$-9 = y$$

36.
$$\frac{3}{p-6} - \frac{4}{p+6} = \frac{48}{p^2-36}$$

$$(p+6)(p-6)\frac{3}{p-6} - (p+6)(p-6)\frac{4}{p+6} = (p+6)(p-6)\frac{48}{(p+6)(p-}$$

$$(p+6)3 - (p-6)4 = 48$$

$$3p + 18 - 4p + 24 = 48$$

$$-p + 42 = 48$$

$$-p = 6$$

$$p = -6$$

But −6 does not check because it makes two denominators in the original equation equal 0. Therefore, there is no solution.

38.
$$\frac{2}{q^2+q-12} = \frac{1}{q+4} + \frac{1}{q-3}$$

$$(q+4)(q-3)\frac{2}{(q+4)(q-3)} = (q+4)(q-3)\frac{1}{q+4} + (q+4)(q-3)\frac{1}{q-3}$$

$$2 = (q-3)1 + (q+4)1$$

$$2 = 2q + 1$$

$$1 = 2q$$

$$\frac{1}{2} = q$$

40.
$$\frac{5}{6r - 12} - \frac{1}{r - 2} = \frac{1}{4r + 12}$$

$$12(r - 2)(r + 3)[\frac{5}{6(r - 2)} - \frac{1}{r - 2}] = 12(r - 2)(r + 3)\frac{1}{4(r + 3)}$$

$$2(r + 3)5 - 12(r + 3) = 3(r - 2)$$

$$10r + 30 - 12r - 36 = 3r - 6$$

$$-2r - 6 = 3r - 6$$

$$-2r = 3r$$

$$0 = 5r$$

$$0 = r$$

42.
$$\frac{m + 7}{m^2 - 3m - 10} + \frac{1}{5m + 10} = 0$$

$$\frac{m + 7}{(m - 5)(m + 2)} + \frac{1}{5(m + 2)} = 0$$

$$5(m - 5)(m + 2)[\frac{m + 7}{(m - 5)(m + 2)} + \frac{1}{5(m + 2)}] = 5(m - 5)(m + 2) \cdot 0$$

$$5(m + 7) + (m - 5)1 = 0$$

$$5m + 35 + m - 5 = 0$$

$$6m + 30 = 0$$

$$6m = -30$$

$$m = -5$$

44.
$$\frac{u - 3}{u + 6} - \frac{u + 1}{u} = \frac{14}{u^2 + 6u}$$

$$u(u + 6)[\frac{u - 3}{u + 6} - \frac{u + 1}{u}] = u(u + 6)\frac{14}{u(u + 6)}$$

$$u(u - 3) - (u + 6)(u + 1) = 14$$

$$u^2 - 3u - (u^2 + 7u + 6) = 14$$

$$-10u - 6 = 14$$

$$-10u = 20$$

$$u = -2$$

46. $\dfrac{PV}{T} = \dfrac{pv}{t}$

$\dfrac{T}{P} \cdot \dfrac{PV}{T} = \dfrac{T}{P} \cdot \dfrac{pv}{t}$

$V = \dfrac{Tpv}{Pt}$

48. $S = \dfrac{a}{1 + r}$

$(1 + r)S = \dfrac{a}{1 + r}(1 + r)$

$S + rS = a$

$rS = a - S$

$r = \dfrac{a - S}{r}$

50. $\dfrac{1}{f} = \dfrac{1}{d_1} + \dfrac{1}{d_2}$

$fd_1d_2 \cdot \dfrac{1}{f} = fd_1d_2 \cdot \dfrac{1}{d_1} + fd_1d_2 \cdot \dfrac{1}{d_2}$

$d_1d_2 = fd_2 + fd_1$

$d_1d_2 - fd_1 = fd_2$

$d_1(d_2 - f) = fd_2$

$d_1 = \dfrac{fd_2}{d_2 - f}$

52. $x = \dfrac{1}{y} - 5$

$y \cdot x = y \cdot \dfrac{1}{y} - y \cdot 5$

$yx = 1 - 5y$

$5y + yx = 1$

$y(5 + x) = 1$

$y = \dfrac{1}{5 + x}$

54. $d = \dfrac{r}{1 - rt}$

$(1 - rt)d = (1 - rt)\dfrac{r}{1 - rt}$

$d - drt = r$

$d = r + drt$

$d = r(1 + dt)$

$\dfrac{d}{1 + dt} = r$

56.
$$y = \frac{x - 1}{x + 1}$$

$$(x + 1)y = (x + 1)\frac{x - 1}{x + 1}$$

$$xy + y = x - 1$$

$$y(x + 1) = x - 1$$

$$y = \frac{x - 1}{x + 1}$$

58.
$$7 - \frac{1}{x} = 5$$

$$x \cdot 7 - x \cdot \frac{1}{x} = x \cdot 5$$

$$7x - 1 = 5x$$

$$-1 = -2x$$

$$\frac{1}{2} = x$$

60.
$$\frac{8.77}{4.73x} + 0.281 = \frac{9.63}{5.09x}$$

$$(4.73)(5.09)x[\frac{8.77}{4.73x} + 0.281] = (4.73)(5.09)x\frac{9.63}{5.09x}$$

$$(5.09)(8.77) + (4.73)(5.09)(0.281)x = (4.73)(9.63)$$

$$44.6393 + 6.7652717x = 45.5499$$

$$6.7652717x = 0.9106$$

$$x \approx 0.135$$

Problem Set 4.6, pp. 174–175

2.
$$\frac{1}{f} = \frac{1}{d_0} + \frac{1}{d_i}$$

$$\frac{1}{f} = \frac{1}{15} + \frac{1}{10}$$

$$30f \cdot \frac{1}{f} = 30f \cdot \frac{1}{15} + 30f \cdot \frac{1}{10}$$

$$30 = 2f + 3f$$

$$30 = 5f$$

$$f = 6 \text{ cm}$$

4. $$\frac{1}{R} = \frac{1}{R_1} + \frac{1}{R_2}$$

$$\frac{1}{2} = \frac{1}{R_1} + \frac{1}{7}$$

$$14R_1 \cdot \frac{1}{2} = 14R_1 \cdot \frac{1}{R_1} + 14R_1 \cdot \frac{1}{7}$$

$$7R_1 = 14 + 2R_1$$

$$5R_1 = 14$$

$$R_1 = 2.8 \text{ ohm}$$

6.   x = the number

$$\frac{9 - x}{17 - x} = \frac{3}{7}$$

$$(9 - x)7 = (17 - x)3$$

$$63 - 7x = 51 - 3x$$

$$63 = 51 + 4x$$

$$12 = 4x$$

$$3 = x$$

8.   x = first number
     5x = second number

$$\frac{1}{x} + \frac{1}{5x} = \frac{2}{15}$$

$$15x\left(\frac{1}{x} + \frac{1}{5x}\right) = 15x \cdot \frac{2}{15}$$

$$15 + 3 = 2x$$

$$18 = 2x$$

$$x = 9$$

$$5x = 45$$

10.   x = first integer
        x + 2 = second integer

$$\frac{1}{x} + \frac{1}{x + 2} = \frac{10}{x(x + 2)}$$

$$x(x + 2)\left[\frac{1}{x} + \frac{1}{x + 2}\right] = x(x + 2)\frac{10}{x(x + 2)}$$

$$(x + 2) + x = 10$$

$$2 + 2x = 10$$

$$2x = 8$$

$$x = 4$$

$$x + 2 = 6$$

12. t = time to plow field together

Portion of field plowed by John in 1 hour	+	portion of field plowed by father in 1 hour	=	Portion of field plowed by both in 1 hour
$\frac{1}{7}$	+	$\frac{1}{5}$	=	$\frac{1}{t}$

$$35t\left(\frac{1}{7} + \frac{1}{5}\right) = 35t \cdot \frac{1}{t}$$

$$5t + 7t = 35$$

$$12t = 35$$

$$t = 2\frac{11}{12}\text{ hr}$$

14. t = time to fill pool if both pipes are open and pump is on

Portion of pool filled by first pipe in 1 day	+	portion of pool filled by second pipe in 1 day	+	portion of pool filled by pump in 1 day	=	portion of pool filled by all in 1 day
$\frac{1}{3}$	+	$\frac{1}{6}$	+	$\left(-\frac{1}{4}\right)$	=	$\frac{1}{t}$

$$12t\left(\frac{1}{3} + \frac{1}{6} - \frac{1}{4}\right) = 12t \cdot \frac{1}{t}$$

$$4t + 2t - 3t = 12$$

$$3t = 12$$

$$t = 4\text{ days}$$

16. s = speed of wind

Time against wind = time with wind

$$\frac{2}{80 - s} = \frac{3}{80 + s}$$

$$2(80 + s) = (80 - s)3$$

$$160 + 2s = 240 - 3s$$

$$2s = 80 - 3s$$

$$5s = 80$$

$$s = 16\text{ mph}$$

18.   x = speed of car
     x + 150 = speed of plane

Time of plane = time of car

$$\frac{840}{x + 150} = \frac{240}{x}$$

$$840x = (x + 150)240$$

$$840x = 240x + 36,000$$

$$600x = 36,000$$

$$x = 60 \text{ mph}$$

$$x + 150 = 210 \text{ mph}$$

20. $\frac{1}{I} = \frac{1}{E} - \frac{1}{Y}$

$$IEY \cdot \frac{1}{I} = IEY\left(\frac{1}{E} - \frac{1}{Y}\right)$$

$$EY = IY - IE$$

$$EY = I(Y - E)$$

$$\frac{EY}{Y - E} = I$$

$$I = \frac{(365.25)(779.87)}{779.87 - 365.25}$$

$$I \approx 687 \text{ days}$$

Problem Set 4.7, pp. 180-181

2. $\frac{10x^2 + 15x - 20}{5} = \frac{10x^2}{5} + \frac{15x}{5} - \frac{20}{5} = 2x^2 + 3x - 4$

4. $\frac{8x^2 - 16x + 12}{4x} = \frac{8x^2}{4x} - \frac{16x}{4x} + \frac{12}{4x} = 2x - 4 + \frac{3}{x}$

6. $(y^5 + y^3 - y^2) \div y^2 = \frac{y^5}{y^2} + \frac{y^3}{y^2} - \frac{y^2}{y^2} = y^3 + y - 1$

8. $\frac{16a^5 - 8a^4 + 10a^3 - 24a}{8a^3} = \frac{16a^5}{8a^3} - \frac{8a^4}{8a^3} + \frac{10a^3}{8a^3} - \frac{24a}{8a^3}$

$$= 2a^2 - a + \frac{5}{4} - \frac{3}{a^2}$$

10. $\dfrac{50p^2q^2 - 75pq + 5q}{25pq^2} = \dfrac{50p^2q^2}{25pq^2} - \dfrac{75pq}{25pq^2} + \dfrac{5q}{25pq^2}$

$$= 2p - \dfrac{3}{q} + \dfrac{1}{5pq}$$

12. $\dfrac{80x^3y^4z^6 + 16x^3y^3z^5 + 48x^2y^4z^4}{16xyz} = \dfrac{80x^3y^4z^6}{16xyz} + \dfrac{16x^3y^3z^5}{16xyz} + \dfrac{48x^2y^4z^4}{16xyz}$

$$= 5x^2y^3z^5 + x^2y^2z^4 + 3xy^3z^3$$

14. $\dfrac{48r^6s^6 + 20r^3s^4 - 12r^3s^3}{12r^3s^3} = \dfrac{48r^6s^6}{12r^3s^3} + \dfrac{20r^3s^4}{12r^3s^3} - \dfrac{12r^3s^3}{12r^3s^3}$

$$= 4r^3s^3 + \dfrac{5}{3}s - 1$$

16. a) $\dfrac{3x - 15}{x - 5} = \dfrac{3(x - 5)}{x - 5} = 3$

    b) $x - 5 \overline{)\, 3x - 15}$ Answer: 3

with quotient 3 above, and below:

$$\underline{3x - 15}$$
$$0$$

18. a) $\dfrac{3m^2 + m - 10}{m + 2} = \dfrac{(3m - 5)(m + 2)}{m + 2} = 3m - 5$

    b) quotient $3m - 5$, $m + 2 \overline{)\, 3m^2 + m - 10}$ Answer: $3m - 5$

$$\underline{3m^2 + 6m}$$
$$-5m - 10$$
$$\underline{-5m - 10}$$
$$0$$

20. a) $\dfrac{8r^2 + 14r - 15}{4r - 3} = \dfrac{(2r + 5)(4r - 3)}{4r - 3} = 2r + 5$

    b) quotient $2r + 5$, $4r - 3 \overline{)\, 8r^2 + 14r - 15}$ Answer: $2r + 5$

$$\underline{8r^2 - 6r}$$
$$20r - 15$$
$$\underline{20r - 15}$$
$$0$$

22.
$$
\begin{array}{r}
x^2 + x + 3 \\
x + 1 \overline{) x^3 + 2x^2 + 4x + 3} \\
\underline{x^3 + x^2} \\
x^2 + 4x \\
\underline{x^2 + x} \\
3x + 3 \\
\underline{3x + 3} \\
0
\end{array}
$$

Answer: $x^2 + x + 3$

24.
$$
\begin{array}{r}
3y^2 + y + 2 \\
3y - 2 \overline{) 9y^3 - 3y^2 + 4y + 4} \\
\underline{9y^3 - 6y^2} \\
3y^2 + 4y \\
\underline{3y^2 - 2y} \\
6y + 4 \\
\underline{6y - 4} \\
8
\end{array}
$$

Answer: $3y^2 + y + 2 + \dfrac{8}{3y - 2}$

26.
$$
\begin{array}{r}
x^2 - x - 4 \\
5x + 2 \overline{) 5x^3 - 3x^2 - 22x + 8} \\
\underline{5x^3 + 2x^2} \\
-5x^2 - 22x \\
\underline{-5x^2 - 2x} \\
-20x + 8 \\
\underline{-20x - 8} \\
16
\end{array}
$$

Answer: $x^2 - x - 4 + \dfrac{16}{5x + 2}$

28.
$$
\begin{array}{r}
6p^2 - 3p + 5 \\
2p + 1 \overline{) 12p^3 + 0p^2 + 7p + 13} \\
\underline{12p^3 + 6p^2} \\
-6p^2 + 7p \\
\underline{-6p^2 - 3p} \\
10p + 13 \\
\underline{10p + 5} \\
8
\end{array}
$$

Answer: $6p^2 - 3p + 5 + \dfrac{8}{2p + 1}$

30.

$$\begin{array}{r} t^2 + 0t + 2 \\ 8t - 7 \overline{)8t^3 - 7t^2 + 16t - 14} \end{array}$$
$$8t^3 - 7t^2$$
$$0t^2 + 16t$$
$$0t^2 - 0t$$
$$16t - 14$$
$$16t - 14$$
$$0$$

Answer: $t^2 + 2$

32.

$$\begin{array}{r} 3r + 4 \\ r^2 + r - 2 \overline{)3r^3 + 7r^2 - r - 8} \end{array}$$
$$3r^3 + 3r^2 - 6r$$
$$4r^2 + 5r - 8$$
$$4r^2 + 4r - 8$$
$$r$$

Answer: $3r + 4 + \dfrac{r}{r^2 + r - 2}$

34.

$$\begin{array}{r} 2s^2 + 6s + 15 \\ 2s^2 - 3s + 5 \overline{)4s^4 + 6s^3 + 22s^2 - 15s + 75} \end{array}$$
$$4s^4 - 6s^3 + 10s^2$$
$$12s^3 + 12s^2 - 15s$$
$$12s^3 - 18s^2 + 30s$$
$$30s^2 - 45s + 75$$
$$30s^2 - 45s + 75$$
$$0$$

Answer: $2s^2 + 6s + 15$

36.

$$\begin{array}{r} 2w^2 + 4w + 3 \\ 4w^2 - 3w \overline{)8w^4 + 10w^3 + 0w^2 + 0w - 7} \end{array}$$
$$8w^4 - 6w^3$$
$$16w^3 + 0w^2$$
$$16w^3 - 12w^2$$
$$12w^2 + 0w$$
$$12w^2 - 9w$$
$$9w - 7$$

Answer: $2w^2 + 4w + 3 + \dfrac{9w - 7}{4w^2 - 3w}$

38.
$$\require{enclose}
\begin{array}{r}
4m \quad - 13 \\[-2pt]
3m^2 + 8 \enclose{longdiv}{12m^3 - 39m^2 + 32m - 104}
\end{array}$$

$$12m^3 \qquad\qquad + 32m$$
$$\underline{}$$
$$-39m^2 + \;\; 0m - 104$$
$$-39m^2 \qquad\quad - 104$$
$$\underline{}$$
$$0$$

Answer: $4m - 13$

40.
$$\begin{array}{r}
x^3 - 2x^2 + 4x \;\; - 8 \\[-2pt]
x + 2 \enclose{longdiv}{x^4 + 0x^3 + 0x^2 + 0x - 16}
\end{array}$$

$$x^4 + 2x^3$$
$$\underline{}$$
$$-2x^3 + 0x^2$$
$$-2x^3 - 4x^2$$
$$\underline{}$$
$$4x^2 + 0x$$
$$4x^2 + 8x$$
$$\underline{}$$
$$-8x - 16$$
$$-8x - 16$$
$$\underline{}$$
$$0$$

Answer: $x^3 - 2x^2 + 4x - 8$

42. $AC = \dfrac{C}{x} = \dfrac{0.01x^2 + 4x + 2500}{x} = 0.01x + 4 + \dfrac{2500}{x}$

44.
$$\begin{array}{r}
x^2 + 2x \;\; + \;\; 5 \\[-2pt]
3x - 1 \enclose{longdiv}{3x^3 + 5x^2 + 13x + k}
\end{array}$$

$$3x^3 - \;\; x^2$$
$$\underline{}$$
$$6x^2 + 13x$$
$$6x^2 - \;\; 2x$$
$$\underline{}$$
$$15x + k$$
$$15x - 5$$

Choose $k = -5$ to make the remainder 0.

46.
$$\begin{array}{r}
5.7r^2 \quad + \;\; 2.2r \quad - 1.85 \\[-2pt]
6.24r + 1.38 \enclose{longdiv}{35.568r^3 + 21.594r^2 - 8.508r - 7.143}
\end{array}$$

$$35.568r^3 + \;\; 7.866r^2$$
$$\underline{}$$
$$13.728r^2 - 8.508r$$
$$13.728r^2 + 3.036r$$
$$\underline{}$$
$$-11.544r - 7.143$$
$$-11.544r - 2.553$$
$$\underline{}$$
$$-4.59$$

Answer: $5.7r^2 + 2.2r - 1.85 - \dfrac{4.59}{6.24r + 1.38}$

Problem Set 4.8, pp. 184–185

2. $\underline{3}$ 2   4   1
$$\underline{\quad\ 6\ \ \ 30\quad}$$
2 10 $\lfloor 31$

$2x + 10 + \dfrac{31}{x - 3}$

4. $\underline{-3}$ 2   4   1
$$\underline{\quad\ -6\ \ \ 6\quad}$$
2 -2 $\lfloor 7$

$2x - 2 + \dfrac{7}{x + 3}$

6. $\underline{3}$ 1   2   -15
$$\underline{\quad\ 3\ \ \ 15\quad}$$
1   5 $\lfloor 0$

$x + 5$

8. $\underline{2}$ 3   -1   -2   -7
$$\underline{\quad\ 6\ \ \ 10\ \ \ 16\quad}$$
3   5   8 $\lfloor 9$

$3x^2 + 5x + 8 + \dfrac{9}{x - 2}$

10. $\underline{5}$ 2   -7   -6   -25
$$\underline{\quad\ 10\ \ \ 15\ \ \ 45\quad}$$
2   3   9 $\lfloor 20$

$2x^2 + 3x + 9 + \dfrac{20}{x - 5}$

12. $\underline{-2}$ 1   4   -1   7
$$\underline{\quad\ -2\ \ \ -4\ \ \ 10\quad}$$
1   2   -5 $\lfloor 17$

$x^2 + 2x - 5 + \dfrac{17}{x + 2}$

14. $\underline{1}$ 1   1   0   -5
$$\underline{\quad\ 1\ \ \ 2\ \ \ 2\quad}$$
1   2   2 $\lfloor -3$

$y^2 + 2y + 2 - \dfrac{3}{y - 1}$

16. $\underline{-1}$ 5   3   0   -1
$$\underline{\quad\ -5\ \ \ 2\ \ \ -2\quad}$$
5   -2   2 $\lfloor -3$

$5m^2 - 2m + 2 - \dfrac{3}{m + 1}$

18. $\underline{-3}$ 4   12   2   11
$$\underline{\quad\ -12\ \ \ 0\ \ \ -6\quad}$$
4   0   2 $\lfloor 5$

$4x^2 + 2 + \dfrac{5}{x + 3}$

20. $\underline{5}$ 1   -1   -15   0
$$\underline{\quad\ 5\ \ \ 20\ \ \ 25\quad}$$
1   4   5 $\lfloor 25$

$r^2 + 4r + 5 + \dfrac{25}{r - 5}$

22. $\underline{2|}$ 1   -3   6   -4   -1

       $\underline{2\ \ -2\ \ \ 8\ \ \ \ 8}$

       1   -1   4   4 $\underline{|7}$

   $p^3 - p^2 + 4p + 4 + \dfrac{7}{p-2}$

24. $\underline{-10|}$ 3   25   -49   12   20

       $\underline{\ \ -30\ \ \ \ 50\ \ -10\ \ -20}$

       3   -5   1   2 $\underline{|0}$

   $3t^3 - 5t^2 + t + 2$

26. $\underline{-4|}$ 1   6   11   0   -59

       $\underline{\ \ -4\ \ -8\ \ -12\ \ \ \ 48}$

       1   2   3   -12 $\underline{|-11}$

   $x^3 + 2x^2 + 3x - 12 - \dfrac{11}{x+4}$

28. $\underline{-1|}$ 3   -2   0   1   1

       $\underline{\ \ -3\ \ \ 5\ \ -5\ \ \ 4}$

       3   -5   5   -4 $\underline{|5}$

   $3y^3 - 5y^2 + 5y - 4 + \dfrac{5}{y+1}$

30. $\underline{-5|}$ 1   6   4   -5   2

       $\underline{\ \ -5\ \ -5\ \ \ 5\ \ \ 0}$

       1   1   -1   0 $\underline{|2}$

   $x^3 + x^2 - x + \dfrac{2}{x+5}$

32. $\underline{-3|}$ 1   -2   -3   1   0   -9

       $\underline{\ \ -3\ \ \ 15\ \ -36\ \ 105\ \ -315}$

       1   -5   12   -35   105 $\underline{|-324}$

   $m^4 - 5m^3 + 12m^2 - 35m + 105 - \dfrac{324}{m+3}$

34. $\underline{2|}$ 1   0   0   -8

       $\underline{\ \ 2\ \ \ 4\ \ \ 8}$

       1   2   4 $\underline{|0}$

   $x^2 + 2x + 4$

36. $\underline{-1|}$ 1   0   0   0   0   1

       $\underline{\ \ -1\ \ \ 1\ \ -1\ \ \ 1\ \ -1}$

       1   -1   1   -1   1 $\underline{|0}$

   $x^4 - x^3 + x^2 - x + 1$

38. $\frac{1}{2}$⌋ 8   0   2   -2
              4   2   2
          8   4   4 ⌊0

$8y^2 + 4y + 4$

40. -2.5⌋ 3.4   7.3   -5.6   -4.8
                 -8.5   3   6.5
           3.4   -1.2   -2.6 ⌊1.7

$3.4x^2 - 1.2x - 2.6 + \dfrac{1.7}{x + 2.5}$

NOTES

CHAPTER 5

RATIONAL EXPONENTS, RADICALS, AND COMPLEX NUMBERS

Problem Set 5.1, pp. 194–195

2. $6^{-1} = \dfrac{1}{6^1} = \dfrac{1}{6}$

4. $5^{-2} = \dfrac{1}{5^2} = \dfrac{1}{25}$

6. $(-3)^{-3} = \dfrac{1}{(-3)^3} = -\dfrac{1}{27}$

8. $(-3)^{-4} = \dfrac{1}{(-3)^4} = \dfrac{1}{81}$

10. $(-9)^{-1} = \dfrac{1}{-9} = -\dfrac{1}{9}$

12. $3 \cdot 5^{-1} = 3 \cdot \dfrac{1}{5} = \dfrac{3}{5}$

14. $2^{-1} + 5^{-1} = \dfrac{1}{2} + \dfrac{1}{5} = \dfrac{5}{10} + \dfrac{2}{10} = \dfrac{7}{10}$

16. $8x^{-2} = 8 \cdot \dfrac{1}{x^2} = \dfrac{8}{x^2}$ and $(8x)^{-2} = \dfrac{1}{(8x)^2} = \dfrac{1}{64x^2}$

18. $\dfrac{1}{4^{-3}} = 4^3 = 64$ and $\dfrac{3}{4^{-3}} = 3 \cdot 4^3 = 3 \cdot 64 = 192$

20. $\dfrac{2}{5r^{-2}} = \dfrac{2}{5} \cdot \dfrac{1}{r^{-2}} = \dfrac{2}{5}r^2$ and $\dfrac{2}{(5r)^{-2}} = 2(5r)^2 = 2(25r^2) = 50r^2$

22. $(\tfrac{2}{5})^{-2} = (\tfrac{5}{2})^2 = \dfrac{25}{4}$ and $(-\tfrac{1}{3})^{-3} = (-\tfrac{3}{1})^3 = -27$

24. $\dfrac{1}{3^{-1}} = 3^1 = 3$ and $3^{-1} = \dfrac{1}{3^1} = \dfrac{1}{3}$

26. $x^{13} \cdot x^{-4} = x^{13 + (-4)} = x^9$

28. $s^{-17} \cdot s \cdot s^9 = s^{-17 + 1 + 9} = s^{-7} = \dfrac{1}{s^7}$

30. $10^{-5} \cdot 10^8 = 10^{-5 + 8} = 10^3$

32. $\dfrac{m^{-9}}{m} = m^{-9 - 1} = m^{-10} = \dfrac{1}{m^{10}}$

34. $\dfrac{m^{-4}}{m^{-8}} = m^{-4 - (-8)} = m^{-4 + 8} = m^4$

36. $\dfrac{y^4}{y^{-7}} = y^{4 - (-7)} = y^{4 + 7} = y^{11}$

38. $\dfrac{10^{-14}}{10^{-1}} = 10^{-14 - (-1)} = 10^{-14 + 1} = 10^{-13} = \dfrac{1}{10^{13}}$

40. $(p^{-2})^9 = p^{(-2)9} = p^{-18} = \dfrac{1}{p^{18}}$

42. $(p^3 q^{-1})^{-6} = p^{3(-6)} q^{(-1)(-6)} = p^{-18} q^6 = \dfrac{q^6}{p^{18}}$

44. $\left(\dfrac{p^2}{q^4}\right)^{-15} = \left(\dfrac{q^4}{p^2}\right)^{15} = \dfrac{(q^4)^{15}}{(p^2)^{15}} = \dfrac{q^{4(15)}}{p^{2(15)}} = \dfrac{q^{60}}{p^{30}}$

46. $(-3u^{-4} v^2)^2 = (-3)^2 (u^{-4})^2 (v^2)^2 = 9u^{-8} v^4 = \dfrac{9v^4}{u^8}$

48. $(5a^{-3} b^2)^{-3} = 5^{-3} (a^{-3})^{-3} (b^2)^{-3} = \dfrac{1}{5^3} a^9 b^{-6} = \dfrac{a^9}{125 b^6}$

50. $\dfrac{x^{-3} y^5}{x^2 y^{-4}} = x^{-3 - 2} y^{5 - (-4)} = x^{-5} y^9 = \dfrac{y^9}{x^5}$

52. $\left(\dfrac{2x}{y^{-3}}\right)^{-2} = \left(\dfrac{y^{-3}}{2x}\right)^2 = \dfrac{(y^{-3})^2}{(2x)^2} = \dfrac{y^{-6}}{4x^2} = \dfrac{1}{4x^2 y^6}$

54. $\dfrac{3a^{-1}}{4b^{-1}} = \dfrac{3}{4} \cdot \dfrac{a^{-1}}{1} \cdot \dfrac{1}{b^{-1}} = \dfrac{3}{4} \cdot \dfrac{1}{a} \cdot b = \dfrac{3b}{4a}$

56. $\dfrac{ax^{-3}}{by^{-4}} = \dfrac{a}{b} \cdot \dfrac{x^{-3}}{1} \cdot \dfrac{1}{y^{-4}} = \dfrac{a}{b} \cdot \dfrac{1}{x^3} \cdot y^4 = \dfrac{ay^4}{bx^3}$

58. $\dfrac{(2a^{-3})^2}{a^{-6}} = \dfrac{2^2(a^{-3})^2}{a^{-6}} = \dfrac{4a^{-6}}{a^{-6}} = 4a^{-6-(-6)} = 4a^0 = 4$

60. $\left(\dfrac{4x^{-5}y^3}{5z^4}\right)^{-2} = \left(\dfrac{5z^4}{4x^{-5}y^3}\right)^2 = \dfrac{5^2(z^4)^2}{4^2(x^{-5})^2(y^3)^2} = \dfrac{25z^8}{16x^{-10}y^6}$

$$= \dfrac{25x^{10}z^8}{16y^6}$$

62. $(-3x^{-5}y^2)^4(xy^{-2})^{-1} = (-3)^4(x^{-5})^4(y^2)^4 x^{-1}(y^{-2})^{-1}$

$$= 81x^{-20}y^8 x^{-1}y^2$$

$$= 81x^{-21}y^{10}$$

$$= \dfrac{81y^{10}}{x^{21}}$$

64. $\left(\dfrac{a^6 b^{-14}}{c^{-3}}\right)^2\left(\dfrac{3a^{-2}}{7b^{-2}}\right) = \dfrac{(a^6)^2(b^{-14})^2}{(c^{-3})^2} \cdot \dfrac{3b^2}{7a^2}$

$$= \dfrac{a^{12}b^{-28}}{c^{-6}} \cdot \dfrac{3b^2}{7a^2}$$

$$= \dfrac{3a^{12}b^{-26}}{7a^2 c^{-6}}$$

$$= \dfrac{3a^{10}c^6}{7b^{26}}$$

66. $(a-b)^{-2} = \dfrac{1}{(a-b)^2}$

68. $c^{-2} + c = \dfrac{1}{c^2} + \dfrac{c}{1} = \dfrac{1}{c^2} + \dfrac{c^3}{c^2} = \dfrac{1+c^3}{c^2}$

70. $\dfrac{d^{-1}+1}{d^{-1}-1} = \dfrac{\frac{1}{d}+1}{\frac{1}{d}-1} \cdot \dfrac{d}{d} = \dfrac{\frac{1}{d}\cdot d + 1\cdot d}{\frac{1}{d}\cdot d - 1\cdot d} = \dfrac{1+d}{1-d}$

72. $\dfrac{t^{-2}+1}{t^{-1}} = \dfrac{\frac{1}{t^2}+1}{\frac{1}{t}} \cdot \dfrac{t^2}{t^2} = \dfrac{\frac{1}{t^2}\cdot t^2 + 1\cdot t^2}{\frac{1}{t}\cdot t^2} = \dfrac{1+t^2}{t}$

74. $\dfrac{r-s}{r^{-2}-s^{-2}} = \dfrac{r-s}{\frac{1}{r^2}-\frac{1}{s^2}} \cdot \dfrac{r^2 s^2}{r^2 s^2} = \dfrac{(r-s)r^2 s^2}{\frac{1}{r^2}\cdot r^2 s^2 - \frac{1}{s^2}\cdot r^2 s^2}$

$= \dfrac{(r-s)r^2 s^2}{s^2 - r^2}$

$= \dfrac{(r-s)r^2 s^2}{(s-r)(s+r)}$

$= -\dfrac{r^2 s^2}{s+r}$

76. $(x^{-2}+y^{-2})^{-1} = \dfrac{1}{x^{-2}+y^{-2}} = \dfrac{1}{\frac{1}{x^2}+\frac{1}{y^2}}\cdot\dfrac{x^2 y^2}{x^2 y^2}$

$= \dfrac{1\cdot x^2 y^2}{\frac{1}{x^2}\cdot x^2 y^2 + \frac{1}{y^2}\cdot x^2 y^2}$

$= \dfrac{x^2 y^2}{y^2 + x^2}$

78. $P = V(1+r)^{-t}$

$P = 25{,}000(1+0.1325)^{-8}$

$P = 25{,}000(1.1325)^{-8}$

$P = \dfrac{25{,}000}{(1.1325)^8}$

$P \approx \$9239.20$

Problem Set 5.2, pp. 198-200

2. $3 \times 10^5 = 300,000$ 4. $8.8 \times 10^6 = 8,800,000$

6. $7 \times 10^{-2} = 0.07$ 8. $2.4 \times 10^{-6} = 0.0000024$

10. $-6.43 \times 10^9 = -6,430,000,000$

12. $-1.15 \times 10^{-7} = -0.000000115$

14. $5.003 \times 10^0 = 5.003$

16. $4.118 \times 10 = 41.18$ 18. $7.07 \times 10^{-1} = 0.707$

20. $9,000,000 = 9 \times 10^6$ 22. $250,000 = 2.5 \times 10^5$

24. $0.003 = 3 \times 10^{-3}$ 26. $0.00047 = 4.7 \times 10^{-4}$

28. $701,000,000,000 = 7.01 \times 10^{11}$

30. $0.00000000005068 = 5.068 \times 10^{-11}$

32. $100,000,000 = 10^8$ 34. $0.000001 = 10^{-6}$

36. $4.41 = 4.41$ 38. $-834.5 = -8.345 \times 10^2$

40. $62.09 = 6.209 \times 10$ 42. $0.331 = 3.31 \times 10^{-1}$

44. $5 \times 10^{20} = 500,000,000,000,000,000,000$

46. $1.7 \times 10^{-24} = 0.0000000000000000000000017$

48. $(4000)(22,000) = (4 \times 10^3)(2.2 \times 10^4)$

$\qquad = (4 \times 2.2)(10^3 \times 10^4)$

$\qquad = 8.8 \times 10^7$

50. $(20,000,000,000)(0.000000003) = (2 \times 10^{10})(3 \times 10^{-9})$

$$= (2 \times 3)(10^{10} \times 10^{-9})$$

$$= 6 \times 10$$

52. $(0.000006)(5,000,000) = (6 \times 10^{-6})(5 \times 10^{6})$

$$= (6 \times 5)(10^{-6} \times 10^{6})$$

$$= 30 \times 10^{0}$$

$$= 30$$

$$= 3 \times 10$$

54. $\dfrac{360,000}{3000} = \dfrac{3.6 \times 10^{5}}{3 \times 10^{3}} = \dfrac{3.6}{3} \times \dfrac{10^{5}}{10^{3}} = 1.2 \times 10^{2}$

56. $\dfrac{0.00008}{0.00000004} = \dfrac{8 \times 10^{-5}}{4 \times 10^{-8}} = \dfrac{8}{4} \times \dfrac{10^{-5}}{10^{-8}} = 2 \times 10^{-5 - (-8)}$

$$= 2 \times 10^{3}$$

58. $\dfrac{0.00000035}{5,000,000,000} = \dfrac{3.5 \times 10^{-7}}{5 \times 10^{9}} = \dfrac{3.5}{5} \times 10^{-7 - 9} = 0.7 \times 10^{-16}$

$$= 7 \times 10^{-1} \times 10^{-16}$$

$$= 7 \times 10^{-17}$$

60. $\dfrac{(520,000)(27,000)}{0.000000000006} = \dfrac{(5.2 \times 10^{5})(2.7 \times 10^{4})}{6 \times 10^{-12}}$

$$= \dfrac{(5.2)(2.7)}{6} \times \dfrac{10^{5}10^{4}}{10^{-12}}$$

$$= 2.34 \times 10^{9 - (-12)}$$

$$= 2.34 \times 10^{21}$$

62. $\dfrac{(6,000,000,000)\,(0.065)}{(0.002)\,(130,000,000)} = \dfrac{(6 \times 10^{9})\,(6.5 \times 10^{-2})}{(2 \times 10^{-3})\,(1.3 \times 10^{8})}$

$$= \dfrac{(6)\,(6.5)}{(2)\,(1.3)} \times \dfrac{10^{9}10^{-2}}{10^{-3}10^{8}}$$

$$= (3)\,(5) \times \dfrac{10^{7}}{10^{5}}$$

$$= 15 \times 10^{2}$$

$$= 1.5 \times 10 \times 10^{2}$$

$$= 1.5 \times 10^{3}$$

64. $\dfrac{(8,500,000,000)\,(0.00078)}{500,000} = \dfrac{(8.5 \times 10^{9})\,(7.8 \times 10^{-4})}{5 \times 10^{5}}$

$$= \dfrac{(8.5)\,(7.8)}{5} \times \dfrac{10^{9}10^{-4}}{10^{5}}$$

$$= (1.7)\,(7.8) \times \dfrac{10^{5}}{10^{5}}$$

$$= 13.26 \times 1$$

$$= 1.326 \times 10$$

66. $\dfrac{(0.00000005)\,(44,000,000)}{(80,000,000,000)\,(0.5)} = \dfrac{(5 \times 10^{-8})\,(4.4 \times 10^{7})}{(8 \times 10^{10})\,(5 \times 10^{-1})}$

$$= \dfrac{(5)\,(4.4)}{(8)\,(5)} \times \dfrac{10^{-8}10^{7}}{10^{10}10^{-1}}$$

$$= \dfrac{4.4}{8} \times \dfrac{10^{-1}}{10^{9}}$$

$$= 0.55 \times 10^{-1 - 9}$$

$$= 0.55 \times 10^{-10}$$

$$= 5.5 \times 10^{-1} \times 10^{-10}$$

$$= 5.5 \times 10^{-11}$$

68. $(8.5 \times 10^{6} \ \frac{calc}{sec}) \ (60 \ \frac{sec}{min}) \ (60 \ \frac{min}{hr}) \ (10 \ hr)$

$= (8.5 \times 10^{6}) \ (6 \times 10) \ (6 \times 10) \ (10)$ calculations

$= (8.5 \times 6 \times 6) \ (10^{6} \times 10 \times 10 \times 10)$ calculations

$= 306 \times 10^{9}$ calculations

$= 3.06 \times 10^{2} \times 10^{9}$ calculations

$= 3.06 \times 10^{11}$ calculations

70. $(1.9 \times 10^{36} \ grams) \ (1.1 \times 10^{-6} \ \frac{tons}{gram})$

$= (1.9 \times 1.1) \ (10^{36} \times 10^{-6})$ tons

$= 2.09 \times 10^{30}$ tons

72. $t = \dfrac{d}{r}$

$t = \dfrac{279,000 \ mi}{186,000 \ mi/sec} = \dfrac{2.79 \times 10^{5}}{1.86 \times 10^{5}} \ \dfrac{mi}{1} \cdot \dfrac{sec}{mi} = 1.5 \ sec$

74. $\dfrac{1,960,000,000,000 \ dollars}{80,000,000 \ taxpayers} = \dfrac{1.96 \times 10^{12}}{8 \times 10^{7}}$ \$/taxpayer

$= 0.245 \times 10^{5}$ \$/taxpayer

$= 24,500$ \$/taxpayer

Problem Set 5.3, pp. 203 - 204

2. $\sqrt{49} = 7$

4. $-\sqrt{49} = -7$

6. $\pm\sqrt{49} = \pm7$

8. $\sqrt{-49}$ is not a real number.

10. $\sqrt[3]{64} = 4$

12. $-\sqrt[3]{64} = -4$

14. $\sqrt[3]{-64} = -4$

16. $-\sqrt[3]{-64} = -(-4) = 4$

18. $\sqrt[4]{625} = 5$

20. $-\sqrt[4]{625} = -5$

22. $\sqrt[4]{-625}$ is not a real number.

24. $\sqrt[5]{32} = 2$

26. $\sqrt[5]{-32} = -2$

28. $\sqrt{\dfrac{9}{49}} = \dfrac{3}{7}$

30. $\pm\sqrt{\dfrac{1}{36}} = \pm\dfrac{1}{6}$

32. $-\sqrt{1} = -1$

34. $-\sqrt{-1}$ is not a real number.

36. $\sqrt{-0} = \sqrt{0} = 0$

38. $\sqrt{1.44} = 1.2$

40. $\sqrt{7^2} = 7$

42. $\sqrt{9^2} = 9$

44. $\sqrt{(-9)^2} = \sqrt{81} = 9$

46. $\sqrt[4]{x^4} = x$

48. $\sqrt[9]{x^9} = x$

50. $\sqrt{x^{10}} = x^5$

52. $\sqrt[5]{x^{10}} = x^2$

54. $\sqrt[3]{x^{15}} = x^5$

56. $\sqrt{81x^2} = 9x$

58. $\sqrt[3]{125x^3} = 5x$

60. $\sqrt[4]{16y^{12}} = 2y^3$

62. $\sqrt[5]{243p^{30}} = 3p^6$

64. $\sqrt{225b^6c^{12}} = 15b^3c^6$

66. $\sqrt[3]{r^{15}s^{90}t^3} = r^5s^{30}t$

68. $\sqrt[3]{-64a^{12}b^6} = -4a^4b^2$

70. $-\sqrt[3]{-512a^{36}b^{99}} = -(-8a^{12}b^{33}) = 8a^{12}b^{33}$

72. $\sqrt[3]{\dfrac{8}{125}p^9q^9} = \dfrac{2}{5}p^3q^3$

74. $\sqrt[11]{(s+t)^{55}} = (s+t)^5$

76. $\sqrt[5]{(v+w)^{15}} = (v+w)^3$

78. $\sqrt{r^2 + 16r + 64} = \sqrt{(r+8)^2} = r + 8$

80. $\sqrt{p^2 + 6pq + 9q^2} = \sqrt{(p+3q)^2} = p + 3q$

82. $\sqrt{36r^{8n}s^{4m}} = 6r^{4n}s^{2m}$

84. $t = 2\pi \sqrt{\dfrac{\ell}{32}}$

$t = 2\pi \sqrt{\dfrac{8}{32}} = 2\pi \sqrt{\dfrac{1}{4}} = 2\pi \left(\dfrac{1}{2}\right) = \pi \text{ sec}$

86. $s = \dfrac{1}{2}(11.8 + 15.5 + 13.4) = \dfrac{1}{2}(40.7) = 20.35$

$A = \sqrt{s(s-a)(s-b)(s-c)}$

$A = \sqrt{20.35(20.35 - 11.8)(20.35 - 15.5)(20.35 - 13.4)}$

$A = \sqrt{20.35(8.55)(4.85)(6.95)}$

$A \approx \sqrt{5864.85}$

$A \approx 76.6 \text{ sq m}$

Problem Set 5.4, pp. 208 - 209

2. $36^{1/2} = \sqrt{36} = 6$ 4. $-36^{1/2} = -\sqrt{36} = -6$

6. $(-36)^{1/2} = \sqrt{-36}$, which is not a real number.

8. $64^{1/3} = \sqrt[3]{64} = 4$ 10. $(-64)^{1/3} = \sqrt[3]{-64} = -4$

12. $16^{1/4} = \sqrt[4]{16} = 2$ 14. $(-1)^{1/5} = \sqrt[5]{-1} = -1$

16. $0^{1/7} = \sqrt[7]{0} = 0$ 18. $\left(\dfrac{25}{144}\right)^{1/2} = \sqrt{\dfrac{25}{144}} = \dfrac{5}{12}$

20. $\left(-\dfrac{1}{1000}\right)^{1/3} = \sqrt[3]{-\dfrac{1}{1000}} = -\dfrac{1}{10}$ 22. $(7x)^{1/2} = \sqrt{7x}$

24. $7x^{1/2} = 7\sqrt{x}$ 26. $(x+5)^{1/3} = \sqrt[3]{x+5}$

28. $(x^2+9)^{1/2} = \sqrt{x^2+9}$ 30. $125^{2/3} = (\sqrt[3]{125})^2 = 5^2 = 25$

32. $\left(\dfrac{9}{16}\right)^{3/2} = \left(\sqrt{\dfrac{9}{16}}\right)^3 = \left(\dfrac{3}{4}\right)^3 = \dfrac{27}{64}$

34. $(-27)^{4/3} = (\sqrt[3]{-27})^4 = (-3)^4 = 81$

36. $(-27)^{2/3} = (\sqrt[3]{-27})^2 = (-3)^2 = 9$

38. $y^{2/5} = \sqrt[5]{y^2}$

40. $15y^{3/5} = 15\sqrt[5]{y^3}$

42. $(5rs)^{2/3} = \sqrt[3]{(5rs)^2} = \sqrt[3]{25r^2s^2}$

44. $(r + 4)^{2/3} = \sqrt[3]{(r + 4)^2} = \sqrt[3]{r^2 + 8r + 16}$

46. $121^{-1/2} = \dfrac{1}{121^{1/2}} = \dfrac{1}{\sqrt{121}} = \dfrac{1}{11}$

48. $9^{-3/2} = \dfrac{1}{9^{3/2}} = \dfrac{1}{(\sqrt{9})^3} = \dfrac{1}{3^3} = \dfrac{1}{27}$

50. $256^{-3/4} = \dfrac{1}{256^{3/4}} = \dfrac{1}{(\sqrt[4]{256})^3} = \dfrac{1}{4^3} = \dfrac{1}{64}$

52. $(-27)^{-2/3} = \dfrac{1}{(-27)^{2/3}} = \dfrac{1}{(\sqrt[3]{-27})^2} = \dfrac{1}{(-3)^2} = \dfrac{1}{9}$

54. $(-1000)^{-5/3} = \dfrac{1}{(-1000)^{5/3}} = \dfrac{1}{(\sqrt[3]{-1000})^5} = \dfrac{1}{(-10)^5} = -\dfrac{1}{100,000}$

56. $\left(\dfrac{16}{25}\right)^{-1/2} = \left(\dfrac{25}{16}\right)^{1/2} = \sqrt{\dfrac{25}{16}} = \dfrac{5}{4}$

58. $x^{1/3} \cdot x^{2/3} = x^{1/3 + 2/3} = x^{3/3} = x^1 = x$

60. $x^{1/6} \cdot x^{7/6} = x^{1/6 + 7/6} = x^{8/6} = x^{4/3}$

62. $\dfrac{y^{3/5}}{y^{1/5}} = y^{3/5 - 1/5} = y^{2/5}$

64. $\dfrac{y^{7/2}}{y^{1/2}} = y^{7/2 - 1/2} = y^{6/2} = y^3$

66. $\dfrac{p^{-3/2} \cdot p^{5/2}}{p^{1/3}} = \dfrac{p^{-3/2 + 5/2}}{p^{1/3}} = \dfrac{p}{p^{1/3}} = p^{1 - 1/3} = p^{2/3}$

68. $\dfrac{p^{-3/2}}{p^{-3/4} \cdot p^{1/4}} = \dfrac{p^{-3/2}}{p^{-3/4 + 1/4}} = \dfrac{p^{-3/2}}{p^{-1/2}} = p^{-3/2 - (-1/2)} = p^{-1} = \dfrac{1}{p}$

70. $(r^{-2/3})^6 = r^{(-2/3)6} = r^{-4} = \dfrac{1}{r^4}$

72. $(r^2 s^{-5})^{1/10} = r^{2(1/10)} s^{-5(1/10)} = r^{1/5} s^{-1/2} = \dfrac{r^{1/5}}{s^{1/2}}$

74. $(3r^{1/2} s^{3/4})^4 = 3^4 r^{(1/2)4} s^{(3/4)4} = 81 r^2 s^3$

76. $(27 p^9 q^6)^{2/3} = 27^{2/3} p^{9(2/3)} q^{6(2/3)} = 9 p^6 q^4$

78. $(\dfrac{m^{20}}{n^{25}})^{1/5} = \dfrac{m^{20(1/5)}}{n^{25(1/5)}} = \dfrac{m^4}{n^5}$

80. $(\dfrac{a^{3/2} b^{5/4}}{c^{-1/2}})^4 = \dfrac{a^{(3/2)4} b^{(5/4)4}}{c^{(-1/2)4}} = \dfrac{a^6 b^5}{c^{-2}} = a^6 b^5 c^2$

82. $(\dfrac{-8a^9 b^{-3}}{c^{18}})^{1/3} = \dfrac{(-8)^{1/3} a^{9(1/3)} b^{-3(1/3)}}{c^{18(1/3)}} = \dfrac{-2a^3 b^{-1}}{c^6} = -\dfrac{2a^3}{bc^6}$

84. $(\dfrac{81 p^{-20}}{q^8})^{-1/4} = (\dfrac{q^8}{81 p^{-20}})^{1/4} = \dfrac{q^{8(1/4)}}{81^{1/4} p^{-20(1/4)}} = \dfrac{q^2}{3p^{-5}} = \dfrac{p^5 q^2}{3}$

86. $(x^{5/2} y^{-5/3})^6 (x^2 y^{-1/2})^{-1} = x^{(5/2)6} y^{(-5/3)6} x^{2(-1)} y^{(-1/2)(-1)}$

$= x^{15} y^{-10} x^{-2} y^{1/2}$

$= x^{13} y^{-19/2}$

$= \dfrac{x^{13}}{y^{19/2}}$

88. $w^{3/2}(w^{1/2} - w) = w^{3/2} \cdot w^{1/2} - w^{3/2} \cdot w = w^2 - w^{5/2}$

90. $y^{1/2}(y^{7/2} - y) = y^{1/2} \cdot y^{7/2} - y^{1/2} \cdot y = y^4 - y^{3/2}$

92. $z^{-3/4}(z^{7/4} + z^{3/4}) = z^{-3/4} \cdot z^{7/4} + z^{-3/4} \cdot z^{3/4}$

$= z^1 + z^0$

$= z + 1$

94. $P = 48x^{1/3}y^{2/3}$

$P = 48 \cdot 20^{1/3} \cdot 20^{2/3}$

$P = 48 \cdot 20^1$

$P = 960$ units

96. $y^{2/3} + y^{-1/3} = y^{-1/3} \cdot y^{3/3} + y^{-1/3} \cdot 1 = y^{-1/3}(y + 1)$

98. $T = d^{3/2}$

$T = (5.201)^{3/2} = (\sqrt{5.201})^3 \approx 11.86$ yr

Problem Set 5.5, pp. 213-214

2. $\sqrt{18} = \sqrt{9}\,\sqrt{2} = 3\sqrt{2}$

4. $\sqrt{48} = \sqrt{16}\,\sqrt{3} = 4\sqrt{3}$

6. $\sqrt{700} = \sqrt{100}\,\sqrt{7} = 10\sqrt{7}$

8. $\sqrt{10}$ cannot be simplified.

10. $\sqrt[3]{200} = \sqrt[3]{8}\,\sqrt[3]{25} = 2\sqrt[3]{25}$

12. $\sqrt[4]{80} = \sqrt[4]{16}\,\sqrt[4]{5} = 2\sqrt[4]{5}$

14. $\sqrt{63x^2} = \sqrt{9x^2}\,\sqrt{7} = 3x\sqrt{7}$

16. $3\sqrt{121y^3} = 3\sqrt{121y^2}\sqrt{y} = 3 \cdot 11y\sqrt{y} = 33y\sqrt{y}$

18. $\sqrt[3]{1000p^7q^6} = \sqrt[3]{1000p^6q^6}\,\sqrt[3]{p} = 10p^2q^2\sqrt[3]{p}$

20. $\sqrt[4]{162m^{19}} = \sqrt[4]{81m^{16}}\ \sqrt[4]{2m^3} = 3m^4\ \sqrt[4]{2m^3}$

22. $\sqrt{\dfrac{12}{25}} = \dfrac{\sqrt{12}}{\sqrt{25}} = \dfrac{\sqrt{4}\ \sqrt{3}}{5} = \dfrac{2\sqrt{3}}{5}$

24. $-\sqrt{\dfrac{300}{196}} = -\dfrac{\sqrt{300}}{\sqrt{196}} = -\dfrac{\sqrt{100}\ \sqrt{3}}{14} = -\dfrac{10\sqrt{3}}{14} = -\dfrac{5\sqrt{3}}{7}$

26. $\sqrt{\dfrac{81x}{100}} = \dfrac{\sqrt{81x}}{\sqrt{100}} = \dfrac{\sqrt{81}\ \sqrt{x}}{10} = \dfrac{9\sqrt{x}}{10}$

28. $\sqrt{\dfrac{64r^3}{81}} = \dfrac{\sqrt{64r^3}}{\sqrt{81}} = \dfrac{\sqrt{64r^2}\ \sqrt{r}}{9} = \dfrac{8r\sqrt{r}}{9}$

30. $\sqrt[3]{\dfrac{8t^8}{125}} = \dfrac{\sqrt[3]{8t^8}}{\sqrt[3]{125}} = \dfrac{\sqrt[3]{8t^6}\ \sqrt[3]{t^2}}{5} = \dfrac{2t^2\ \sqrt[3]{t^2}}{5}$

32. $\sqrt[5]{\dfrac{x}{y^{15}}} = \dfrac{\sqrt[5]{x}}{\sqrt[5]{y^{15}}} = \dfrac{\sqrt[5]{x}}{y^3}$

34. $\sqrt{\sqrt{7}} = (7^{1/2})^{1/2} = 7^{1/4} = \sqrt[4]{7}$

36. $\sqrt{\sqrt[4]{x}} = (x^{1/4})^{1/2} = x^{1/8} = \sqrt[8]{x}$

38. $\sqrt{\sqrt{\sqrt{10}}} = ((10^{1/2})^{1/2})^{1/2} = 10^{1/8} = \sqrt[8]{10}$

40. $\sqrt[6]{z^3} = z^{3/6} = z^{1/2} = \sqrt{z}$

42. $\sqrt[4]{49} = 49^{1/4} = (7^2)^{1/4} = 7^{1/2} = \sqrt{7}$

44. $\sqrt[6]{100x^6y^4} = (100x^6y^4)^{1/6} = ((10x^3y^2)^2)^{1/6} = (10x^3y^2)^{1/3} = \sqrt[3]{10x^3y^2}$

46. $\sqrt{44} = \sqrt{4}\ \sqrt{11} = 2\sqrt{11}$

48. $\sqrt{a^6b^5} = \sqrt{a^6b^4}\ \sqrt{b} = a^3b^2\sqrt{b}$

50. $\sqrt{\dfrac{3}{49}} = \dfrac{\sqrt{3}}{\sqrt{49}} = \dfrac{\sqrt{3}}{7}$

52. $-\sqrt{24} = -\sqrt{4}\sqrt{6} = -2\sqrt{6}$

54. $\sqrt[3]{\sqrt[5]{x}} = (x^{1/5})^{1/3} = x^{1/15} = \sqrt[15]{x}$

56. $\sqrt{125} = \sqrt{25}\sqrt{5} = 5\sqrt{5}$

58. $\sqrt[3]{81} = \sqrt[3]{27}\sqrt[3]{3} = 3\sqrt[3]{3}$

60. $\sqrt{\dfrac{30y^2}{49}} = \dfrac{\sqrt{30y^2}}{\sqrt{49}} = \dfrac{\sqrt{y^2}\sqrt{30}}{7} = \dfrac{y\sqrt{30}}{7}$

62. $\sqrt[4]{512} = \sqrt[4]{256}\,\sqrt[4]{2} = 4\sqrt[4]{2}$

64. $2\sqrt{75} = 2\sqrt{25}\sqrt{3} = 2\cdot5\sqrt{3} = 10\sqrt{3}$

66. $\sqrt{c^9 d^{11}} = \sqrt{c^8 d^{10}}\sqrt{cd} = c^4 d^5\sqrt{cd}$

68. $\sqrt[4]{4x^2} = (4x^2)^{1/4} = ((2x)^2)^{1/4} = (2x)^{1/2} = \sqrt{2x}$

70. $\sqrt{\dfrac{x^2 y^7}{9z^4}} = \dfrac{\sqrt{x^2 y^7}}{\sqrt{9z^4}} = \dfrac{\sqrt{x^2 y^6}\sqrt{y}}{3z^2} = \dfrac{xy^3\sqrt{y}}{3z^2}$

72. $\sqrt{\dfrac{pq}{q^5}} = \sqrt{\dfrac{p}{q^4}} = \dfrac{\sqrt{p}}{\sqrt{q^4}} = \dfrac{\sqrt{p}}{q^2}$

74. $\sqrt{288r^3 s^8} = \sqrt{144r^2 s^8}\,\sqrt{2r} = 12rs^4\sqrt{2r}$.

76. $\sqrt[3]{54t^6} = \sqrt[3]{27t^6}\,\sqrt[3]{2} = 3t^2\sqrt[3]{2}$

78. $\sqrt[3]{-125k^{17}} = \sqrt[3]{-125k^{15}}\,\sqrt[3]{k^2} = -5k^5\sqrt[3]{k^2}$

80. $\sqrt[4]{256w^5} = \sqrt[4]{256w^4}\,\sqrt[4]{w} = 4w\sqrt[4]{w}$

82. $\sqrt[3]{\dfrac{81}{64}} = \dfrac{\sqrt[3]{81}}{\sqrt[3]{64}} = \dfrac{\sqrt[3]{27}\,\sqrt[3]{3}}{4} = \dfrac{3\sqrt[3]{3}}{4}$

84. $-\sqrt[4]{\dfrac{m^{25}}{1296}} = -\dfrac{\sqrt[4]{m^{25}}}{\sqrt[4]{1296}} = -\dfrac{\sqrt[4]{m^{24}}\,\sqrt[4]{m}}{6} = -\dfrac{m^6\sqrt[4]{m}}{6}$

86. $\sqrt[6]{27} = (27)^{1/6} = (3^3)^{1/6} = 3^{1/2} = \sqrt{3}$

88. $\sqrt{36\sqrt{k}} = \sqrt{36}\sqrt{\sqrt{k}} = 6(k^{1/2})^{1/2} = 6k^{1/4} = 6\sqrt[4]{k}$

90. $-\sqrt[3]{-27a^{15}b^{24}c^{21}} = -(-3a^5b^8c^7) = 3a^5b^8c^7$

92. $\sqrt{25 - 16} = \sqrt{9} = 3$, but $\sqrt{25} - \sqrt{16} = 5 - 4 = 1.$

94. $d = \sqrt{1.5a}$

  $d = \sqrt{1.5(20,000)} = \sqrt{30,000} = \sqrt{10,000}\sqrt{3} = 100\sqrt{3}$ mi

96. $v = \sqrt{2gR}$

  $v = \sqrt{2\cdot32\ \dfrac{ft}{sec^2}\cdot3960\ mi}$

  $v = \sqrt{2\cdot32\ \dfrac{ft}{sec^2}\cdot3960\ mi\cdot5280\ \dfrac{ft}{mi}}$

  $v = \sqrt{2\cdot32\cdot3960\cdot5280\ \dfrac{ft^2}{sec^2}}$

  $v = \sqrt{(2)(2^5)(2^3\cdot3^2\cdot5\cdot11)(2^5\cdot3\cdot5\cdot11)\ \dfrac{ft^2}{sec^2}}$

  $v = \sqrt{2^{14}\cdot3^3\cdot5^2\cdot11^2\ \dfrac{ft^2}{sec^2}}$

  $v = \sqrt{2^{14}\cdot3^2\cdot5^2\cdot11^2\ \dfrac{ft^2}{sec^2}}\sqrt{3}$

  $v = 2^7\cdot3\cdot5\cdot11\ \dfrac{ft}{sec}\sqrt{3}$

  $v = 21,120\sqrt{3}\ \dfrac{ft}{sec}$

Problem Set 5.6, pp. 217-218

2. $\sqrt{5}\cdot\sqrt{7} = \sqrt{5\cdot7} = \sqrt{35}$

4. $\sqrt{15}\cdot\sqrt{x} = \sqrt{15x}$

6. $\sqrt{3}\cdot\sqrt{3} = \sqrt{3\cdot3} = \sqrt{9} = 3$

8. $\sqrt{10}\cdot\sqrt{10} = \sqrt{100} = 10$

10. $\sqrt{2} \cdot \sqrt{2} \cdot \sqrt{2} = 2\sqrt{2}$

12. $(\sqrt{y})^6 = \sqrt{y} \cdot \sqrt{y} \cdot \sqrt{y} \cdot \sqrt{y} \cdot \sqrt{y} \cdot \sqrt{y} = y \cdot y \cdot y = y^3$

14. $(\sqrt{y})^7 = (\sqrt{y})^6 \sqrt{y} = y^3 \sqrt{y}$

16. $\sqrt{3} \sqrt{5m} = \sqrt{15m}$

18. $\sqrt{2x} \sqrt{10x} = \sqrt{20x^2} = \sqrt{4x^2} \sqrt{5} = 2x\sqrt{5}$

20. $\sqrt{3y} \sqrt{15y} = \sqrt{45y} = \sqrt{9} \sqrt{5y} = 3\sqrt{5y}$

22. $\sqrt{2} \sqrt{2z} \sqrt{11} = \sqrt{44z} = \sqrt{4} \sqrt{11z} = 2\sqrt{11z}$

24. $(3\sqrt{k})^3 = 3^3 (\sqrt{k})^3 = 27k\sqrt{k}$

26. $(\sqrt{x + 4})^2 = x + 4$

28. $\sqrt{r^2 s} \sqrt{rs^5} = \sqrt{r^3 s^6} = \sqrt{r^2 s^6} \sqrt{r} = rs^3 \sqrt{r}$

30. $\sqrt{40} \sqrt{15t} = \sqrt{600t} = \sqrt{100} \sqrt{6t} = 10\sqrt{6t}$

32. $\sqrt{2a^3} \sqrt{5a^3} \sqrt{10a} = \sqrt{100a^7} = \sqrt{100a^6} \sqrt{a} = 10a^3 \sqrt{a}$

34. $\sqrt[3]{9p^2} \sqrt[3]{3p} = \sqrt[3]{27p^3} = 3p$

36. $\sqrt[3]{5} \sqrt[3]{40q} = \sqrt[3]{200q} = \sqrt[3]{8} \sqrt[3]{25q} = 2\sqrt[3]{25q}$

38. $\sqrt[3]{2r^4 s} \sqrt[3]{9r^3 s^3} = \sqrt[3]{18r^7 s^4} = \sqrt[3]{r^6 s^3} \sqrt[3]{18rs} = r^2 s \sqrt[3]{18rs}$

40. $\sqrt[4]{9a} \sqrt[4]{9b} = \sqrt[4]{81ab} = \sqrt[4]{81} \sqrt[4]{ab} = 3\sqrt[4]{ab}$

42. $\sqrt{4.5 \times 10^{-3}} \sqrt{8 \times 10^{11}} = \sqrt{36 \times 10^8} = 6 \times 10^4$

44. $\dfrac{\sqrt{75}}{\sqrt{3}} = \sqrt{\dfrac{75}{3}} = \sqrt{25} = 5$

46. $\dfrac{\sqrt{81x}}{\sqrt{x}} = \sqrt{\dfrac{81x}{x}} = \sqrt{81} = 9$

48. $\dfrac{\sqrt{30}}{\sqrt{5}} = \sqrt{\dfrac{30}{5}} = \sqrt{6}$

50. $\dfrac{\sqrt{288a}}{\sqrt{2}} = \sqrt{\dfrac{288a}{2}} = \sqrt{144a} = \sqrt{144}\sqrt{a} = 12\sqrt{a}$

52. $\dfrac{\sqrt{4y}}{\sqrt{64y^3}} = \sqrt{\dfrac{4y}{64y^3}} = \sqrt{\dfrac{1}{16y^2}} = \dfrac{\sqrt{1}}{\sqrt{16y^2}} = \dfrac{1}{4y}$

54. $\dfrac{\sqrt{54m^5}}{\sqrt{3m}} = \sqrt{\dfrac{54m^5}{3m}} = \sqrt{18m^4} = \sqrt{9m^4}\sqrt{2} = 3m^2\sqrt{2}$

56. $\dfrac{12\sqrt{160c}}{4\sqrt{10}} = \dfrac{12}{4}\sqrt{\dfrac{160c}{10}} = 3\sqrt{16c} = 3\cdot 4\sqrt{c} = 12\sqrt{c}$

58. $\dfrac{\sqrt{91a^5b^9c^3}}{\sqrt{13abc}} = \sqrt{\dfrac{91a^5b^9c^3}{13abc}} = \sqrt{7a^4b^8c^2} = a^2b^4c\sqrt{7}$

60. $\dfrac{\sqrt{768r^7s}}{\sqrt{12r^3s^7}} = \sqrt{\dfrac{768r^7s}{12r^3s^7}} = \sqrt{\dfrac{64r^4}{s^6}} = \dfrac{8r^2}{s^3}$

62. $\dfrac{\sqrt[3]{32t^4}}{\sqrt[3]{4t}} = \sqrt[3]{\dfrac{32t^4}{4t}} = \sqrt[3]{8t^3} = 2t$

64. $\dfrac{\sqrt[4]{108p^5q^5}}{\sqrt[4]{3p^3q^3}} = \sqrt[4]{\dfrac{108p^5q^5}{3p^3q^3}} = \sqrt[4]{36p^2q^2} = \sqrt[4]{(6pq)^2}$

$$= (6pq)^{2/4}$$

$$= (6pq)^{1/2}$$

$$= \sqrt{6pq}$$

66. $\dfrac{\sqrt{3r^3}\sqrt{3r^3s}}{\sqrt{49s^3}} = \dfrac{\sqrt{9r^6s}}{\sqrt{49s^3}} = \sqrt{\dfrac{9r^6s}{49s^3}} = \sqrt{\dfrac{9r^6}{49s^2}} = \dfrac{3r^3}{7s}$

68. $\dfrac{\sqrt{9.9 \times 10^7}}{\sqrt{1.1 \times 10}} = \sqrt{\dfrac{9.9 \times 10^7}{1.1 \times 10}} = \sqrt{9 \times 10^6} = 3 \times 10^3$

70. $\sqrt{5}\,\sqrt[3]{5} = 5^{1/2}\cdot 5^{1/3} = 5^{3/6}\cdot 5^{2/6} = 5^{5/6} = \sqrt[6]{5^5} = \sqrt[6]{3125}$

72. $\sqrt[6]{x}\sqrt{x} = x^{1/6}\cdot x^{1/2} = x^{1/6}\cdot x^{3/6} = x^{4/6} = x^{2/3} = \sqrt[3]{x^2}$

74. $\dfrac{\sqrt{10}}{\sqrt[4]{10}} = \dfrac{10^{1/2}}{10^{1/4}} = \dfrac{10^{2/4}}{10^{1/4}} = 10^{1/4} = \sqrt[4]{10}$

76. $\dfrac{\sqrt{k}}{\sqrt[8]{k}} = \dfrac{k^{1/2}}{k^{1/8}} = \dfrac{k^{4/8}}{k^{1/8}} = k^{3/8} = \sqrt[8]{k^3}$

78. $\sqrt{a} \ \sqrt[6]{a} = a^{1/2} \cdot a^{1/6} = a^{3/6} \cdot a^{1/6} = a^{4/6} = a^{2/3} = \sqrt[3]{a^2}$

80. $(\sqrt{3})^1 = \sqrt{3}$, which is irrational.

$(\sqrt{3})^2 = 3$, which is rational.

82. $\dfrac{\sqrt[4]{20,736}}{\sqrt[4]{1296}} = \dfrac{12}{6} = 2$

$\sqrt[4]{\dfrac{20,736}{1296}} = \sqrt[4]{16} = 2$

Problem Set 5.7, pp. 220–221

2. $2\sqrt{5} + 4\sqrt{5} = (2 + 4)\sqrt{5} = 6\sqrt{5}$

4. $19\sqrt{10} - 7\sqrt{10} = (19 - 7)\sqrt{10} = 12$

6. $3\sqrt{7} + 7\sqrt{3}$ cannot be simplified.

8. $\sqrt[3]{6} + \sqrt[5]{6}$ cannot be simplified.

10. $\sqrt{x} + 5\sqrt{x} = (1 + 5)\sqrt{x} = 6\sqrt{x}$

12. $7\sqrt{5y} - 3\sqrt{5y} = (7 - 3)\sqrt{5y} = 4\sqrt{5y}$

14. $8\sqrt{x} + 2\sqrt{y}$ cannot be simplified.

16. $3\sqrt{13} + 2\sqrt{13} - 8\sqrt{13} = (3 + 2 - 8)\sqrt{13} = -3\sqrt{13}$

18. $9\sqrt{m} - \sqrt{m} + 3m = (9 - 1)\sqrt{m} + 3m = 8\sqrt{m} + 3m$

20. $5a\sqrt{3} + 2a\sqrt{3} = (5a + 2a)\sqrt{3} = 7a\sqrt{3}$

22. $\sqrt{20} + \sqrt{5} = 2\sqrt{5} + \sqrt{5} = 3\sqrt{5}$

24. $\sqrt{12} + \sqrt{48} = 2\sqrt{3} + 4\sqrt{3} = 6\sqrt{3}$

26. $3\sqrt{32} - \sqrt{2} = 3\cdot 4\sqrt{2} - \sqrt{2} = 12\sqrt{2} - \sqrt{2} = 11\sqrt{2}$

28. $3\sqrt{7} - 2\sqrt{63} = 3\sqrt{7} - 2\cdot 3\sqrt{7} = 3\sqrt{7} - 6\sqrt{7} = -3\sqrt{7}$

30. $2\sqrt{175} - \sqrt{28} - \sqrt{63} = 2\cdot 5\sqrt{7} - 2\sqrt{7} - 3\sqrt{7}$

$$= 10\sqrt{7} - 2\sqrt{7} - 3\sqrt{7}$$

$$= 5\sqrt{7}$$

32. $5\sqrt{250} + 6\sqrt{360} + 7\sqrt{490} = 5\cdot 5\sqrt{10} + 6\cdot 6\sqrt{10} + 7\cdot 7\sqrt{10}$

$$= 25\sqrt{10} + 36\sqrt{10} + 49\sqrt{10}$$

$$= 110\sqrt{10}$$

34. $\sqrt{3x} + \sqrt{75x} = \sqrt{3x} + 5\sqrt{3x} = 6\sqrt{3x}$

36. $\sqrt{8a} + 3\sqrt{2a} - 5\sqrt{18a} = 2\sqrt{2a} + 3\sqrt{2a} - 5\cdot 3\sqrt{2a}$

$$= -10\sqrt{2a}$$

38. $2\sqrt{11r^3} + \sqrt{99r^3} = 2r\sqrt{11r} + 3r\sqrt{11r} = 5r\sqrt{11r}$

40. $\sqrt[3]{16} + \sqrt[3]{2} = 2\sqrt[3]{2} + \sqrt[3]{2} = 3\sqrt[3]{2}$

42. $\sqrt[3]{125y} - \sqrt[3]{27y} = 5\sqrt[3]{y} - 3\sqrt[3]{y} = 2\sqrt[3]{y}$

44. $\sqrt[3]{40p} - \sqrt[3]{5p} + \sqrt[3]{625p} = 2\sqrt[3]{5p} - \sqrt[3]{5p} + 5\sqrt[3]{5p} = 6\sqrt[3]{5p}$

46. $2\sqrt[3]{48} - 5\sqrt[3]{162} - 2\sqrt[3]{384} = 2\cdot 2\sqrt[3]{6} - 5\cdot 3\sqrt[3]{6} - 2\cdot 4\sqrt[3]{6}$

$$= 4\sqrt[3]{6} - 15\sqrt[3]{6} - 8\sqrt[3]{6}$$

$$= -19\sqrt[3]{6}$$

48. $\sqrt[4]{81t} + 2\sqrt[4]{t} = 3\sqrt[4]{t} + 2\sqrt[4]{t} = 5\sqrt[4]{t}$

50. $7(\sqrt{2} + 4) = 7\sqrt{2} + 7\cdot 4 = 7\sqrt{2} + 28$

52. $2(4\sqrt{14} - \sqrt{7}) = 2 \cdot 4\sqrt{14} - 2\sqrt{7} = 8\sqrt{14} - 2\sqrt{7}$

54. $\sqrt{5}(\sqrt{3} + \sqrt{7}) = \sqrt{5}\sqrt{3} + \sqrt{5}\sqrt{7} = \sqrt{15} + \sqrt{35}$

56. $\sqrt{3}(2\sqrt{6} + \sqrt{3}) = \sqrt{3} \cdot 2\sqrt{6} + \sqrt{3}\sqrt{3} = 2\sqrt{18} + 3$
$$= 2 \cdot 3\sqrt{2} + 3$$
$$= 6\sqrt{2} + 3$$

58. $\sqrt{x}(\sqrt{x} + \sqrt{xy}) = \sqrt{x}\sqrt{x} + \sqrt{x}\sqrt{xy} = x + \sqrt{x^2 y}$
$$= x + x\sqrt{y}$$

60. $3\sqrt{2}(4\sqrt{6} + 5\sqrt{2}) = 3\sqrt{2} \cdot 4\sqrt{6} + 3\sqrt{2} \cdot 5\sqrt{2} = 12\sqrt{12} + 15 \cdot 2$
$$= 12 \cdot 2\sqrt{3} + 30$$
$$= 24\sqrt{3} + 30$$

62. $\sqrt[3]{3}(\sqrt[3]{9} - \sqrt[3]{4}) = \sqrt[3]{3}\sqrt[3]{9} - \sqrt[3]{3}\sqrt[3]{4} = \sqrt[3]{27} - \sqrt[3]{12}$
$$= 3 - \sqrt[3]{12}$$

64. $(\sqrt{5} + 1)(\sqrt{2} + 7) = \sqrt{5}\sqrt{2} + 7\sqrt{5} + \sqrt{2} + 7 = \sqrt{10} + 7\sqrt{5} + \sqrt{2} + 7$

66. $(\sqrt{7} + \sqrt{2})(\sqrt{7} - \sqrt{2}) = \sqrt{7}\sqrt{7} - \sqrt{7}\sqrt{2} + \sqrt{2}\sqrt{7} - \sqrt{2}\sqrt{2} = 7 - 2 = 5$

68. $(\sqrt{5} - \sqrt{3})(\sqrt{5} - \sqrt{2}) = \sqrt{5}\sqrt{5} - \sqrt{5}\sqrt{2} - \sqrt{3}\sqrt{5} + \sqrt{3}\sqrt{2}$
$$= 5 - \sqrt{10} - \sqrt{15} + \sqrt{6}$$

70. $(\sqrt{13} + \sqrt{5})(\sqrt{2} + \sqrt{3}) = \sqrt{13}\sqrt{2} + \sqrt{13}\sqrt{3} + \sqrt{5}\sqrt{2} + \sqrt{5}\sqrt{3}$
$$= \sqrt{26} + \sqrt{39} + \sqrt{10} + \sqrt{15}$$

72. $(\sqrt{2} + 10\sqrt{7})(\sqrt{2} + \sqrt{3}) = \sqrt{2}\sqrt{2} + \sqrt{2}\sqrt{3} + 10\sqrt{7}\sqrt{2} + 10\sqrt{7}\sqrt{3}$
$$= 2 + \sqrt{6} + 10\sqrt{14} + 10\sqrt{21}$$

74. $(t + \sqrt{7})(t - \sqrt{7}) = t^2 - \sqrt{7}t + \sqrt{7}t - \sqrt{7}\sqrt{7} = t^2 - 7$

76. $(\sqrt{s} - 3)(\sqrt{s} + 3) = \sqrt{s}\sqrt{s} + 3\sqrt{s} - 3\sqrt{s} - 9 = s - 9$

78. $(2\sqrt{6} - 10)(4\sqrt{3} + 1) = 2\sqrt{6}\cdot4\sqrt{3} + 2\sqrt{6} - 10\cdot4\sqrt{3} - 10$

$$= 8\sqrt{18} + 2\sqrt{6} - 40\sqrt{3} - 10$$
$$= 8\cdot3\sqrt{2} + 2\sqrt{6} - 40\sqrt{3} - 10$$
$$= 24\sqrt{2} + 2\sqrt{6} - 40\sqrt{3} - 10$$

80. $(\sqrt{7} + 3)^2 = (\sqrt{7} + 3)(\sqrt{7} + 3)$

$$= \sqrt{7}\sqrt{7} + 3\sqrt{7} + 3\sqrt{7} + 9$$
$$= 7 + 6\sqrt{7} + 9$$
$$= 16 + 6\sqrt{7}$$

82. $(\sqrt{2} + \sqrt{5})^2 = (\sqrt{2} + \sqrt{5})(\sqrt{2} + \sqrt{5})$

$$= \sqrt{2}\sqrt{2} + \sqrt{2}\sqrt{5} + \sqrt{5}\sqrt{2} + \sqrt{5}\sqrt{5}$$
$$= 2 + \sqrt{10} + \sqrt{10} + 5$$
$$= 7 + 2\sqrt{10}$$

84. $(\sqrt{x} + 2\sqrt{y})^2 = (\sqrt{x} + 2\sqrt{y})(\sqrt{x} + 2\sqrt{y})$

$$= \sqrt{x}\sqrt{x} + \sqrt{x}\cdot2\sqrt{y} + 2\sqrt{y}\sqrt{x} + 2\sqrt{y}\cdot2\sqrt{y}$$
$$= x + 2\sqrt{xy} + 2\sqrt{xy} + 4y$$
$$= x + 4\sqrt{xy} + 4y$$

86. $(\sqrt{x + 9} - 1)^2 = (\sqrt{x + 9} - 1)(\sqrt{x + 9} - 1)$

$$= \sqrt{x + 9}\sqrt{x + 9} - \sqrt{x + 9} - \sqrt{x + 9} + 1$$
$$= x + 9 - 2\sqrt{x + 9} + 1$$
$$= x + 10 - 2\sqrt{x + 9}$$

88. $(\sqrt{2x + 3} + \sqrt{x})^2$

$$= (\sqrt{2x + 3} + \sqrt{x})(\sqrt{2x + 3} + \sqrt{x})$$
$$= \sqrt{2x + 3}\sqrt{2x + 3} + \sqrt{2x + 3}\sqrt{x} + \sqrt{x}\sqrt{2x + 3} + \sqrt{x}\sqrt{x}$$
$$= 2x + 3 + \sqrt{2x^2 + 3x} + \sqrt{2x^2 + 3x} + x$$
$$= 3x + 3 + 2\sqrt{2x^2 + 3x}$$

90. $(\sqrt{3} + 1)^3 = (\sqrt{3} + 1)(\sqrt{3} + 1)(\sqrt{3} + 1)$

$\qquad = (\sqrt{3}\sqrt{3} + \sqrt{3} + \sqrt{3} + 1)(\sqrt{3} + 1)$

$\qquad = (3 + 2\sqrt{3} + 1)(\sqrt{3} + 1)$

$\qquad = (4 + 2\sqrt{3})(\sqrt{3} + 1)$

$\qquad = 4\sqrt{3} + 4 + 2\sqrt{3}\sqrt{3} + 2\sqrt{3}$

$\qquad = 4\sqrt{3} + 4 + 2\cdot 3 + 2\sqrt{3}$

$\qquad = 6\sqrt{3} + 10$

92. $(\sqrt{3} + \sqrt{5})^2 \approx (1.732 + 2.236)^2 = (3.968)^2 \approx 15.7$

$\quad (\sqrt{3} + \sqrt{5})^2 = (\sqrt{3} + \sqrt{5})(\sqrt{3} + \sqrt{5})$

$\qquad = \sqrt{3}\sqrt{3} + \sqrt{3}\sqrt{5} + \sqrt{5}\sqrt{3} + \sqrt{5}\sqrt{5}$

$\qquad = 3 + \sqrt{15} + \sqrt{15} + 5$

$\qquad = 8 + 2\sqrt{15}$

$\qquad \approx 8 + 2(3.87)$

$\qquad \approx 15.7$

Problem Set 5.8, pp. 225-226

2. $\dfrac{1}{\sqrt{3}} = \dfrac{1\sqrt{3}}{\sqrt{3}\sqrt{3}} = \dfrac{\sqrt{3}}{3}$

4. $\dfrac{\sqrt{3}}{\sqrt{7}} = \dfrac{\sqrt{3}\sqrt{7}}{\sqrt{7}\sqrt{7}} = \dfrac{\sqrt{21}}{7}$

6. $\dfrac{21}{\sqrt{7}} = \dfrac{21\sqrt{7}}{\sqrt{7}\sqrt{7}} = \dfrac{21\sqrt{7}}{7} = 3\sqrt{7}$

8. $\dfrac{3}{\sqrt{6}} = \dfrac{3\sqrt{6}}{\sqrt{6}\sqrt{6}} = \dfrac{3\sqrt{6}}{6} = \dfrac{\sqrt{6}}{2}$

10. $\dfrac{5}{\sqrt{8}} = \dfrac{5\sqrt{8}}{\sqrt{8}\sqrt{8}} = \dfrac{5\sqrt{8}}{8} = \dfrac{5\cdot 2\sqrt{2}}{8} = \dfrac{10\sqrt{2}}{8} = \dfrac{5\sqrt{2}}{4}$

12. $\dfrac{2\sqrt{5}}{\sqrt{x}} = \dfrac{2\sqrt{5}\;\sqrt{x}}{\sqrt{x}\sqrt{x}} = \dfrac{2\sqrt{5x}}{x}$

14. $\dfrac{6\sqrt{5}}{\sqrt{10}} = \dfrac{6\sqrt{5}\;\sqrt{10}}{\sqrt{10}\sqrt{10}} = \dfrac{6\sqrt{50}}{10} = \dfrac{6\cdot5\sqrt{2}}{10} = \dfrac{30\sqrt{2}}{10} = 3\sqrt{2}$

16. $\dfrac{10}{\sqrt[3]{5}} = \dfrac{10\cdot\sqrt[3]{25}}{\sqrt[3]{5}\;\sqrt[3]{25}} = \dfrac{10\;\sqrt[3]{25}}{\sqrt[3]{125}} = \dfrac{10\;\sqrt[3]{25}}{5} = 2\sqrt[3]{25}$

18. $\dfrac{6}{\sqrt[3]{9}} = \dfrac{6\cdot\sqrt[3]{3}}{\sqrt[3]{9}\;\sqrt[3]{3}} = \dfrac{6\;\sqrt[3]{3}}{\sqrt[3]{27}} = \dfrac{6\;\sqrt[3]{3}}{3} = 2\sqrt[3]{3}$

20. $\sqrt{\dfrac{7}{3}} = \dfrac{\sqrt{7}}{\sqrt{3}} = \dfrac{\sqrt{7}\sqrt{3}}{\sqrt{3}\sqrt{3}} = \dfrac{\sqrt{21}}{3}$

22. $\sqrt{\dfrac{2}{r}} = \dfrac{\sqrt{2}}{\sqrt{r}} = \dfrac{\sqrt{2}\sqrt{r}}{\sqrt{r}\sqrt{r}} = \dfrac{\sqrt{2r}}{r}$

24. $\sqrt{\dfrac{3}{10t}} = \dfrac{\sqrt{3}}{\sqrt{10t}} = \dfrac{\sqrt{3}\sqrt{10t}}{\sqrt{10t}\sqrt{10t}} = \dfrac{\sqrt{30t}}{10t}$

26. $\dfrac{\sqrt{27x^3}}{\sqrt{15xy}} = \sqrt{\dfrac{27x^3}{15xy}} = \sqrt{\dfrac{9x^2}{5y}} = \dfrac{\sqrt{9x^2}}{\sqrt{5y}} = \dfrac{3x}{\sqrt{5y}}$

$\qquad\qquad = \dfrac{3x\sqrt{5y}}{\sqrt{5y}\sqrt{5y}}$

$\qquad\qquad = \dfrac{3x\sqrt{5y}}{5y}$

28. $\dfrac{\sqrt[3]{15}}{\sqrt[3]{20p}} = \sqrt[3]{\dfrac{15}{20p}} = \sqrt[3]{\dfrac{3}{4p}} = \dfrac{\sqrt[3]{3}}{\sqrt[3]{4p}} = \dfrac{\sqrt[3]{3}\cdot\sqrt[3]{2p^2}}{\sqrt[3]{4p}\cdot\sqrt[3]{2p^2}}$

$\qquad\qquad\qquad = \dfrac{\sqrt[3]{6p^2}}{\sqrt[3]{8p^3}}$

$\qquad\qquad\qquad = \dfrac{\sqrt[3]{6p^2}}{2p}$

30. $\dfrac{1}{\sqrt{2} - 1} = \dfrac{1\,(\sqrt{2} + 1)}{(\sqrt{2} - 1)\,(\sqrt{2} + 1)} = \dfrac{\sqrt{2} + 1}{2 - 1} = \sqrt{2} + 1$

32. $\dfrac{1}{2 + \sqrt{3}} = \dfrac{1\,(2 - \sqrt{3})}{(2 + \sqrt{3})\,(2 - \sqrt{3})} = \dfrac{2 - \sqrt{3}}{4 - 3} = 2 - \sqrt{3}$

34. $\dfrac{2}{\sqrt{11} - 3} = \dfrac{2\,(\sqrt{11} + 3)}{(\sqrt{11} - 3)\,(\sqrt{11} + 3)} = \dfrac{2\,(\sqrt{11} + 3)}{11 - 9} = \dfrac{2\,(\sqrt{11} + 3)}{2} = \sqrt{11} + 3$

36. $\dfrac{14}{\sqrt{5} + 1} = \dfrac{14\,(\sqrt{5} - 1)}{(\sqrt{5} + 1)\,(\sqrt{5} - 1)} = \dfrac{14\,(\sqrt{5} - 1)}{5 - 1} = \dfrac{14\,(\sqrt{5} - 1)}{4} = \dfrac{7\,(\sqrt{5} - 1)}{2}$

38. $\dfrac{1}{2\sqrt{7} + 1} = \dfrac{1\,(2\sqrt{7} - 1)}{(2\sqrt{7} + 1)\,(2\sqrt{7} - 1)} = \dfrac{2\sqrt{7} - 1}{4 \cdot 7 - 1} = \dfrac{2\sqrt{7} - 1}{27}$

40. $\dfrac{12}{\sqrt{7} - \sqrt{5}} = \dfrac{12\,(\sqrt{7} + \sqrt{5})}{(\sqrt{7} - \sqrt{5})\,(\sqrt{7} + \sqrt{5})} = \dfrac{12\,(\sqrt{7} + \sqrt{5})}{7 - 5}$

$$= \dfrac{12\,(\sqrt{7} + \sqrt{5})}{2}$$

$$= 6\,(\sqrt{7} + \sqrt{5})$$

42. $\dfrac{\sqrt{2}}{\sqrt{17} + \sqrt{2}} = \dfrac{\sqrt{2}\,(\sqrt{17} - \sqrt{2})}{(\sqrt{17} + \sqrt{2})\,(\sqrt{17} - \sqrt{2})} = \dfrac{\sqrt{34} - 2}{17 - 2} = \dfrac{\sqrt{34} - 2}{15}$

44. $\dfrac{\sqrt{11} + \sqrt{5}}{\sqrt{11} - \sqrt{5}} = \dfrac{(\sqrt{11} + \sqrt{5})\,(\sqrt{11} + \sqrt{5})}{(\sqrt{11} - \sqrt{5})\,(\sqrt{11} + \sqrt{5})} = \dfrac{11 + 2\sqrt{55} + 5}{11 - 5} = \dfrac{16 + 2\sqrt{55}}{6}$

$$= \dfrac{2\,(8 + \sqrt{55})}{6}$$

$$= \dfrac{8 + \sqrt{55}}{3}$$

46. $\dfrac{2 + \sqrt{30}}{\sqrt{5} - \sqrt{6}} = \dfrac{(2 + \sqrt{30})\,(\sqrt{5} + \sqrt{6})}{(\sqrt{5} - \sqrt{6})\,(\sqrt{5} + \sqrt{6})} = \dfrac{2\sqrt{5} + 2\sqrt{6} + \sqrt{150} + \sqrt{180}}{5 - 6}$

$$= \dfrac{2\sqrt{5} + 2\sqrt{6} + 5\sqrt{6} + 6\sqrt{5}}{-1}$$

$$= \dfrac{8\sqrt{5} + 7\sqrt{6}}{-1}$$

$$= -8\sqrt{5} - 7\sqrt{6}$$

48. $\dfrac{3\sqrt{y}}{3\sqrt{x}+5\sqrt{y}} = \dfrac{3\sqrt{y}\,(3\sqrt{x}-5\sqrt{y})}{(3\sqrt{x}+5\sqrt{y})\,(3\sqrt{x}-5\sqrt{y})} = \dfrac{9\sqrt{xy}-15y}{9x-25y}$

50. $\dfrac{m-9}{\sqrt{m}+3} = \dfrac{(m-9)\,(\sqrt{m}-3)}{(\sqrt{m}+3)\,(\sqrt{m}-3)} = \dfrac{(m-9)\,(\sqrt{m}-3)}{m-9} = \sqrt{m}-3$

52. $\dfrac{\sqrt{3}}{2} - \dfrac{1}{\sqrt{3}} = \dfrac{\sqrt{3}}{2} - \dfrac{\sqrt{3}}{3} = \dfrac{3\sqrt{3}}{6} - \dfrac{2\sqrt{3}}{6} = \dfrac{\sqrt{3}}{6}$

54. $\dfrac{1}{\sqrt{7}} + \sqrt{7} = \dfrac{\sqrt{7}}{7} + \dfrac{\sqrt{7}}{1} = \dfrac{\sqrt{7}}{7} + \dfrac{7\sqrt{7}}{7} = \dfrac{8\sqrt{7}}{7}$

56. $\dfrac{\sqrt{8}}{\sqrt{10}} + \dfrac{\sqrt{10}}{\sqrt{8}} = \dfrac{\sqrt{80}}{10} + \dfrac{\sqrt{80}}{8} = \dfrac{4\sqrt{5}}{10} + \dfrac{4\sqrt{5}}{8} = \dfrac{2\sqrt{5}}{5} + \dfrac{2\sqrt{5}}{4} = \dfrac{8\sqrt{5}}{20} + \dfrac{10\sqrt{5}}{20}$

$$= \dfrac{18\sqrt{5}}{20}$$

$$= \dfrac{9\sqrt{5}}{10}$$

58. $\sqrt{48} + \sqrt{\dfrac{1}{3}} = 4\sqrt{3} + \dfrac{\sqrt{1}}{\sqrt{3}} = \dfrac{4\sqrt{3}}{1} + \dfrac{1}{\sqrt{3}} = \dfrac{4\sqrt{3}}{1} + \dfrac{\sqrt{3}}{3} = \dfrac{12\sqrt{3}}{3} + \dfrac{\sqrt{3}}{3} = \dfrac{13\sqrt{3}}{3}$

60. $\sqrt{60m} + \sqrt{\dfrac{3m}{5}} = 2\sqrt{15m} + \dfrac{\sqrt{3m}}{\sqrt{5}} = \dfrac{2\sqrt{15m}}{1} + \dfrac{\sqrt{15m}}{5} = \dfrac{10\sqrt{15m}}{5} + \dfrac{\sqrt{15m}}{5} = \dfrac{11\sqrt{15m}}{5}$

62. $\dfrac{1}{\sqrt{x}} - \dfrac{1}{\sqrt{y}} = \dfrac{\sqrt{x}}{x} - \dfrac{\sqrt{y}}{y} = \dfrac{y\sqrt{x}}{xy} - \dfrac{x\sqrt{y}}{xy} = \dfrac{y\sqrt{x}-x\sqrt{y}}{xy}$

64. $\dfrac{\sqrt{5}+\sqrt{2}}{\sqrt{5}-\sqrt{2}} \approx \dfrac{2.236+1.414}{2.236-1.414} = \dfrac{3.65}{0.822} \approx 4.44$

$\dfrac{\sqrt{5}+\sqrt{2}}{\sqrt{5}-\sqrt{2}} = \dfrac{(\sqrt{5}+\sqrt{2})\,(\sqrt{5}+\sqrt{2})}{(\sqrt{5}-\sqrt{2})\,(\sqrt{5}+\sqrt{2})}$

$$= \dfrac{5+2\sqrt{10}+2}{5-2}$$

$$= \dfrac{7+2\sqrt{10}}{3}$$

$$\approx \dfrac{7+2(3.162)}{3}$$

$$\approx 4.44$$

Problem Set 5.9, pp. 230-231

2. $\sqrt{-25} = \sqrt{25}\sqrt{-1} = 5i$

4. $\sqrt{-64} = \sqrt{64}\sqrt{-1} = 8i$

6. $\sqrt{-\frac{1}{16}} = \sqrt{\frac{1}{16}}\sqrt{-1} = \frac{1}{4}i$

8. $\sqrt{-3} = \sqrt{3}\sqrt{-1} = \sqrt{3}i$

10. $\sqrt{-18} = \sqrt{18}\sqrt{-1} = 3\sqrt{2}i$

12. $6\sqrt{-81} = 6\sqrt{81}\sqrt{-1} = 6 \cdot 9i = 54i$

14. $-\sqrt{-121} = -\sqrt{121}\sqrt{-1} = -11i$

16. $\sqrt{-\frac{5}{9}} = \sqrt{\frac{5}{9}}\sqrt{-1} = \frac{\sqrt{5}}{3}i$

18. a) True   b) True   c) True

20. $(8 + 3i) + (5 + 9i) = 8 + 3i + 5 + 9i = 13 + 12i$

22. $(-7 + 9i) + (3 - 4i) = -7 + 9i + 3 - 4i = -4 + 5i$

24. $(6 + 3i) + (-2 - 3i) = 6 + 3i - 2 - 3i = 4$

26. $(-8 - 14i) + (8 - 7i) = -8 - 14i + 8 - 7i = -21i$

28. $(-5 - i) + 6i = -5 - i + 6i = -5 + 5i$

30. $(17 + i) + 11 = 17 + i + 11 = 28 + i$

32. $8i + 2i = 10i$

34. $(7 + 5i) - (2 + 3i) = 7 + 5i - 2 - 3i = 5 + 2i$

36. $(-8 - 4i) - (-8 - 4i) = -8 - 4i + 8 + 4i = 0$

38. $13 - (2 + 7i) = 13 - 2 - 7i = 11 - 7i$

40. $(-5 - 6i) - (-15) = -5 - 6i + 15 = 10 - 6i$

42. i − 19i = −18i

44. $(-4 - 3i) - [(6 + 9i) - (7 - i)] = (-4 - 3i) - [6 + 9i - 7 + i]$

$$= (-4 - 3i) - [-1 + 10i]$$
$$= -4 - 3i + 1 - 10i$$
$$= -3 - 13i$$

46. $(2 + 6i)(4 + 5i) = 8 + 10i + 24i + 30i^2$

$$= 8 + 34i + 30(-1)$$
$$= -22 + 34i$$

48. $(4 + 2i)(8 - 3i) = 32 - 12i + 16i - 6i^2$

$$= 32 + 4i - 6(-1)$$
$$= 38 + 4i$$

50. $(-7 + i)(2 - 5i) = -14 + 35i + 2i - 5i^2$

$$= -14 + 37i - 5(-1)$$
$$= -9 + 37i$$

52. $(-8 - 6i)(-1 - 4i) = 8 + 32i + 6i + 24i^2$

$$= 8 + 38i + 24(-1)$$
$$= -16 + 38i$$

54. $(6 - 3i)(6 + 3i) = 36 + 18i - 18i - 9i^2$

$$= 36 - 9(-1)$$
$$= 45$$

56. $(-5i)(-2i) = 10i^2 = 10(-1) = -10$

58. $4i(3 - 2i) = 12i - 8i^2 = 12i - 8(-1) = 8 + 12i$

60. $7(-3 + 2i) = -21 + 14i$

62. $(5 - 3i)^2 = (5 - 3i)(5 - 3i) = 25 - 15i - 15i + 9i^2$

$$= 25 - 30i + 9(-1)$$

$$= 16 - 30i$$

64. $(1 + \sqrt{3}i)^2 = (1 + \sqrt{3}i)(1 + \sqrt{3}i) = 1 + \sqrt{3}i + \sqrt{3}i + 3i^2$

$$= 1 + 2\sqrt{3}i + 3(-1)$$

$$= -2 + 2\sqrt{3}i$$

66. $(1 - i)^3 = (1 - i)(1 - i)(1 - i) = (1 - i - i + i^2)(1 - i)$

$$= (1 - 2i + (-1))(1 - i)$$

$$= -2i(1 - i)$$

$$= -2i + 2i^2$$

$$= -2i + 2(-1)$$

$$= -2 - 2i$$

68. $\dfrac{6}{1 + i} = \dfrac{6(1 - i)}{(1 + i)(1 - i)} = \dfrac{6 - 6i}{1 - i^2} = \dfrac{6 - 6i}{1 - (-1)} = \dfrac{6 - 6i}{2} = 3 - 3i$

70. $\dfrac{10i}{1 + 3i} = \dfrac{10i(1 - 3i)}{(1 + 3i)(1 - 3i)} = \dfrac{10i - 30i^2}{1 - 9i^2} = \dfrac{10i - 30(-1)}{1 - 9(-1)}$

$$= \dfrac{30 + 10i}{10}$$

$$= 3 + i$$

72. $\dfrac{5 + 3i}{1 - i} = \dfrac{(5 + 3i)(1 + i)}{(1 - i)(1 + i)} = \dfrac{5 + 8i + 3i^2}{1 - i^2} = \dfrac{5 + 8i + 3(-1)}{1 - (-1)}$

$$= \dfrac{2 + 8i}{2}$$

$$= 1 + 4i$$

74. $\dfrac{10 + 10i}{3 + i} = \dfrac{(10 + 10i)(3 - i)}{(3 + i)(3 - i)} = \dfrac{30 + 20i - 10i^2}{9 - i^2}$

$$= \dfrac{30 + 20i - 10(-1)}{9 - (-1)}$$

$$= \dfrac{40 + 20i}{10}$$

$$= 4 + 2i$$

76. $\dfrac{2 + 5i}{4 - 3i} = \dfrac{(2 + 5i)(4 + 3i)}{(4 - 3i)(4 + 3i)} = \dfrac{8 + 26i + 15i^2}{16 - 9i^2} = \dfrac{8 + 26i + 15(-1)}{16 - 9(-1)}$

$$= \dfrac{-7 + 26i}{25}$$

$$= -\dfrac{7}{25} + \dfrac{26}{25}i$$

78. $\dfrac{5i}{-3 + 4i} = \dfrac{5i(-3 - 4i)}{(-3 + 4i)(-3 - 4i)} = \dfrac{-15i - 20i^2}{9 - 16i^2} = \dfrac{-15i - 20(-1)}{9 - 16(-1)}$

$$= \dfrac{20 - 15i}{25}$$

$$= \dfrac{20}{25} - \dfrac{15}{25}i$$

$$= \dfrac{4}{5} - \dfrac{3}{5}i$$

80. $\dfrac{1 + 4i}{2i} = \dfrac{(1 + 4i)(-2i)}{2i(-2i)} = \dfrac{-2i - 8i^2}{-4i^2} = \dfrac{-2i - 8(-1)}{-4(-1)} = \dfrac{8 - 2i}{4}$

$$= \dfrac{8}{4} - \dfrac{2}{4}i$$

$$= 2 - \dfrac{1}{2}i$$

82. $\dfrac{18i}{-3i} = -6$

84. $\dfrac{-1}{-7i} = \dfrac{-1(7i)}{(-7i)(7i)} = \dfrac{-7i}{-49i^2} = \dfrac{-7i}{-49(-1)} = \dfrac{-7i}{49} = -\dfrac{1}{7}i$

86. $\dfrac{8 - 12i}{4} = \dfrac{8}{4} - \dfrac{12i}{4} = 2 - 3i$

88. $(5i)^2 = 25i^2 = 25(-1) = -25$

    $(-5i)^2 = 25i^2 = 25(-1) = -25$

90. $(a + bi)(a - bi) = a^2 - abi + abi - b^2i^2$

    $\qquad\qquad\qquad = a^2 - b^2(-1)$

    $\qquad\qquad\qquad = a^2 + b^2$

    Since a and b are real numbers, $a^2 + b^2$ is a real number.

92. $\dfrac{1.1 + 9.9i}{3.5 - 0.5i} = \dfrac{(1.1 + 9.9i)(3.5 + 0.5i)}{(3.5 - 0.5i)(3.5 + 0.5i)}$

    $\qquad = \dfrac{3.85 + 35.2i + 4.95i^2}{12.25 - 0.25i^2}$

    $\qquad = \dfrac{3.85 + 35.2i + 4.95(-1)}{12.25 - 0.25(-1)}$

    $\qquad = \dfrac{-1.1 + 35.2i}{12.5}$

    $\qquad = \dfrac{-1.1}{12.5} + \dfrac{35.2}{12.5}i$

    $\qquad = -0.088 + 2.816i$

148

NOTES

CHAPTER 6

SECOND-DEGREE EQUATIONS AND INEQUALITIES

Problem Set 6.1, pp. 240–242

2. $(x - 4)(x + 5) = 0$

$x - 4 = 0$ or $x + 5 = 0$

$x = 4 \qquad x = -5$

4. $(3y + 1)(y + 2) = 0$

$3y + 1 = 0$ or $y + 2 = 0$

$y = -\frac{1}{3} \qquad y = -2$

6. $(z + 3)(z - 3) = 0$

$z + 3 = 0$ or $z - 3 = 0$

$z = -3 \qquad z = 3$

8. $(r - 9)(r - 9) = 0$

$r - 9 = 0$ or $r - 9 = 0$

$r = 9 \qquad r = 9$

10. $6m(m + 7) = 0$

$6m = 0$ or $m + 7 = 0$

$m = 0 \qquad m = -7$

12. $x^2 - 7x + 6 = 0$

$(x - 1)(x - 6) = 0$

$x - 1 = 0$ or $x - 6 = 0$

$x = 1 \qquad x = 6$

14. $x^2 + 2x = 35$

$x^2 + 2x - 35 = 0$

$(x + 7)(x - 5) = 0$

$x + 7 = 0$ or $x - 5 = 0$

$x = -7 \qquad x = 5$

16. $3y^2 + 16y = 12$

$3y^2 + 16y - 12 = 0$

$(3y - 2)(y + 6) = 0$

$3y - 2 = 0$ or $y + 6 = 0$

$y = \frac{2}{3} \qquad y = -6$

18. $y^2 = 12 - 4y$

$y^2 + 4y - 12 = 0$

$(y + 6)(y - 2) = 0$

$y + 6 = 0$ or $y - 2 = 0$

$y = -6 \qquad y = 2$

20. $z^2 + 8z + 16 = 0$

$(z + 4)(z + 4) = 0$

$z + 4 = 0$

$z = -4$

22. $$z^2 = 6z - 9$$
$$z^2 - 6z + 9 = 0$$
$$(z - 3)(z - 3) = 0$$
$$z - 3 = 0$$
$$z = 3$$

24. $$2m^2 + 9m = 5$$
$$2m^2 + 9m - 5 = 0$$
$$(2m - 1)(m + 5) = 0$$
$$2m - 1 = 0 \text{ or } m + 5 = 0$$
$$m = \frac{1}{2} \qquad m = -5$$

26. $$r^2 - 19r = 0$$
$$r(r - 19) = 0$$
$$r = 0 \text{ or } r - 19 = 0$$
$$r = 19$$

28. $$6x^2 = 18x$$
$$6x^2 - 18x = 0$$
$$6x(x - 3) = 0$$
$$6x = 0 \text{ or } x - 3 = 0$$
$$x = 0 \qquad x = 3$$

30. $$6t^2 + 10 = 23t$$
$$6t^2 - 23t + 10 = 0$$
$$(3t - 10)(2t - 1) = 0$$
$$3t - 10 = 0 \text{ or } 2t - 1 = 0$$
$$t = \frac{10}{3} \qquad t = \frac{1}{2}$$

32. $$p^2 = 9$$
$$p = \pm 3$$

34. $$x^2 - 25 = 0$$
$$x^2 = 25$$
$$x = \pm 5$$

36. $$2y^2 - 98 = 0$$
$$2y^2 = 98$$
$$y^2 = 49$$
$$y = \pm 7$$

38. $$25s^2 + 9 = 30s$$
$$25s^2 - 30s + 9 = 0$$
$$(5s - 3)(5s - 3) = 0$$
$$5s - 3 = 0$$
$$s = \frac{3}{5}$$

40. $$4z^2 - 9 = 0$$
$$4z^2 = 9$$
$$z^2 = \frac{9}{4}$$
$$z = \pm \frac{3}{2}$$

42. $8m^2 - 2m = 0$

$2m(4m - 1) = 0$

$2m = 0$ or $4m - 1 = 0$

$m = 0$　　　$m = \dfrac{1}{4}$

44. $r^2 = -r$

$r^2 + r = 0$

$r(r + 1) = 0$

$r = 0$ or $r + 1 = 0$

$r = -1$

46. $7x^2 = 0$

$x^2 = 0$

$x = 0$

48. $y^2 - 2 = 0$

$y^2 = 2$

$y = \pm\sqrt{2}$

50. $5z^2 - 90 = 0$

$5z^2 = 90$

$z^2 = 18$

$z = \pm\sqrt{18}$

$z = \pm 3\sqrt{2}$

52. $(r - 8)(r + 11) = 42$

$r^2 + 3r - 88 = 42$

$r^2 + 3r - 130 = 0$

$(r + 13)(r - 10) = 0$

$r + 13 = 0$ or $r - 10 = 0$

$r = -13$　　　$r = 10$

54. $m(4m + 5) = m - 1$

$4m^2 + 5m = m - 1$

$4m^2 + 4m + 1 = 0$

$(2m + 1)(2m + 1) = 0$

$2m + 1 = 0$

$m = -\dfrac{1}{2}$

56. $(3x + 1)^2 - x^2 = 9x + 1$

$9x^2 + 6x + 1 - x^2 = 9x + 1$

$8x^2 - 3x = 0$

$x(8x - 3) = 0$

$x = 0$ or $8x - 3 = 0$

$x = \dfrac{3}{8}$

58. $(3p + 5)(p + 1) = 4(2p + 3) + 14$

$3p^2 + 8p + 5 = 8p + 12 + 14$

$3p^2 - 21 = 0$

$3p^2 = 21$

$p^2 = 7$

$p = \pm\sqrt{7}$

60. $x^2 - 25a^2 = 0$

$\qquad x^2 = 25a^2$

$\qquad x = \pm 5a$

62. $7x^2 - 21ax = 0$

$\qquad 7x(x - 3a) = 0$

$\qquad 7x = 0 \text{ or } x - 3a = 0$

$\qquad x = 0 \qquad\qquad x = 3a$

64. $x^2 - 8ax + 16a^2 = 0$

$\qquad (x - 4a)(x - 4a) = 0$

$\qquad\qquad x - 4a = 0$

$\qquad\qquad\qquad x = 4a$

66. $\qquad x^2 - 5ax = 6a^2$

$\qquad x^2 - 5ax - 6a^2 = 0$

$\qquad (x - 6a)(x + a) = 0$

$\qquad x - 6a = 0 \text{ or } x + a = 0$

$\qquad\qquad x = 6a \qquad\quad x = -a$

68. $12x^2 + 7ax - 10a^2 = 0$

$\qquad (3x - 2a)(4x + 5a) = 0$

$\qquad 3x - 2a = 0 \text{ or } 4x + 5a = 0$

$\qquad\quad x = \dfrac{2a}{3} \qquad\quad x = -\dfrac{5a}{4}$

70. $\qquad c^2 = a^2 + b^2$

$\qquad c^2 - a^2 = b$

$\qquad \pm\sqrt{c^2 - a^2} = b$

72. $h = 15 \text{ in.} = \dfrac{15}{12} \text{ ft} = \dfrac{5}{4} \text{ ft}$

$\quad s^2 = 64\left(\dfrac{5}{4}\right) = 80$

$\quad s = \sqrt{80} = \sqrt{16 \cdot 5} = 4\sqrt{5} \approx 8.9 \text{ ft/sec}$

74. $\qquad\qquad R = C$

$\qquad 125x - 4x^2 = 25x + 600$

$\qquad\qquad 0 = 4x^2 - 100x + 600$

$\qquad\qquad 0 = x^2 - 25x + 150$

$\qquad\qquad 0 = (x - 10)(x - 15)$

$\qquad x - 10 = 0 \text{ or } x - 15 = 0$

$\qquad\qquad x = 10 \qquad\quad x = 15$

76. $x^2 = 7x + 8$

$x^2 - 7x - 8 = 0$

$(x - 8)(x + 1) = 0$

$x - 8 = 0$ or $x + 1 = 0$

$x = 8$ $x = -1$

78. $x = 4$ or $x = -4$

$x - 4 = 0$ or $x + 4 = 0$

$(x - 4)(x + 4) = 0$

$x^2 - 16 = 0$

80. $x = \frac{2}{3}$ or $x = \frac{2}{3}$

$3x = 2$ or $3x = 2$

$3x - 2 = 0$ or $3x - 2 = 0$

$(3x - 2)(3x - 2) = 0$

$9x^2 - 12x + 4 = 0$

82. $x^2 = 1171$

$x = \sqrt{1171}$

$x \approx 34.22$

Problem Set 6.2, pp. 244-245

2. $(\frac{10}{2})^2 = 5^2 = 25$

4. $(\frac{-4}{2})^2 = (-2)^2 = 4$

6. $(\frac{12}{2})^2 = 6^2 = 36$

8. $(\frac{-5}{2})^2 = \frac{25}{4}$

10. $(\frac{-1}{2})^2 = \frac{1}{4}$

12. $(\frac{1}{2}\cdot\frac{2}{3})^2 = (\frac{1}{3})^2 = \frac{1}{9}$

14. $(x - 4)^2 = 9$

$x - 4 = \pm 3$

$x = 4 \pm 3$

$x = 4 + 3$ or $x = 4 - 3$

$x = 7$ $x = 1$

16. $(x + 5)^2 = 49$

$x + 5 = \pm 7$

$x = -5 \pm 7$

$x = -5 + 7$ or $x = -5 - 7$

$x = 2$ $x = -12$

18. $(y - 2)^2 = 3$

$y - 2 = \pm\sqrt{3}$

$y = 2 \pm\sqrt{3}$

20. $(p + 1)^2 = 8$

$p + 1 = \pm\sqrt{8}$

$p + 1 = \pm 2\sqrt{2}$

$p = -1 \pm 2\sqrt{2}$

22. $(r - \frac{1}{2})^2 = \frac{9}{4}$ 24. $(m + \frac{3}{2})^2 = \frac{7}{4}$

$r - \frac{1}{2} = \pm\frac{3}{2}$ $m + \frac{3}{2} = \pm\frac{\sqrt{7}}{2}$

$r = \frac{1}{2} \pm \frac{3}{2}$ $m = -\frac{3}{2} \pm \frac{\sqrt{7}}{2}$

$r = 2$ or $r = -1$ $m = \frac{-3 \pm \sqrt{7}}{2}$

26.   a) $x^2 - 10x + 21 = 0$ b) $x^2 - 10x + 21 = 0$

$x^2 - 10x = -21$ $(x - 3)(x - 7) = 0$

$x^2 - 20x + 25 = -21 + 25$ $x - 3 = 0$ or $x - 7 = 0$

$(x - 5)^2 = 4$ $x = 3$ $x = 7$

$x - 5 = \pm 2$

$x = 5 \pm 2$

$x = 7$ or $x = 3$

28.   a) $x^2 - 4x - 12 = 0$ b) $x^2 - 4x - 12 = 0$

$x^2 - 4x = 12$ $(x - 6)(x + 2) = 0$

$x^2 - 4x + 4 = 12 + 4$ $x - 6 = 0$ or $x + 2 = 0$

$(x - 2)^2 = 16$ $x = 6$ $x = -2$

$x - 2 = \pm 4$

$x = 2 \pm 4$

$x = 6$ or $x = -2$

30.   a) $x^2 + 6x + 9 = 0$ b) $x^2 + 6x + 9 = 0$

$x^2 + 6x = -9$ $(x + 3)(x + 3) = 0$

$x^2 + 6x + 9 = -9 + 9$ $x + 3 = 0$

$(x + 3)^2 = 0$ $x = -3$

$x + 3 = 0$

$x = -3$

32. a) $\qquad$ $x^2 + 2x = 0$

$x^2 + 2x + 1 = 0 + 1$

$(x + 1)^2 = 1$

$x + 1 = \pm 1$

$x = -1 \pm 1$

$x = 0$ or $x = -2$

b) $\qquad$ $x^2 + 2x = 0$

$x(x + 2) = 0$

$x = 0$ or $x + 2 = 0$

$x = -2$

34. $x^2 - 6x + 4 = 0$

$x^2 - 6x = -4$

$x^2 - 6x + 9 = -4 + 9$

$(x + 3)^2 = 5$

$x + 3 = \pm\sqrt{5}$

$x = -3 \pm\sqrt{5}$

36. $y^2 - 4y - 2 = 0$

$y^2 - 4y = 2$

$y^2 - 4y + 4 = 2 + 4$

$(y - 2)^2 = 6$

$y - 2 = \pm\sqrt{6}$

$y = 2 \pm\sqrt{6}$

38. $z^2 + 10z + 17 = 0$

$z^2 + 10z = -17$

$z^2 + 10z + 25 = -17 + 25$

$(z + 5)^2 = 8$

$z + 5 = \pm\sqrt{8}$

$z + 5 = \pm 2\sqrt{2}$

$z = -5 \pm 2\sqrt{2}$

40. $\qquad$ $p^2 - 8p = 11$

$p^2 - 8p + 16 = 11 + 16$

$(p - 4)^2 = 27$

$p - 4 = \pm\sqrt{27}$

$p - 4 = \pm 3\sqrt{3}$

$p = 4 \pm 3\sqrt{3}$

42. $\qquad$ $r^2 + r = 3$

$r^2 + r + \dfrac{1}{4} = 3 + \dfrac{1}{4}$

$(r + \dfrac{1}{2})^2 = \dfrac{13}{4}$

$r + \dfrac{1}{2} = \pm\dfrac{\sqrt{13}}{2}$

$r = -\dfrac{1}{2} \pm \dfrac{\sqrt{13}}{2}$

44. $2m^2 + 28m + 10 = 0$

$2m^2 + 28m = -10$

$m^2 + 14m = -5$

$m^2 + 14m + 49 = -5 + 49$

$(m + 7)^2 = 45$

$m + 7 = \pm\sqrt{45}$

$m + 7 = \pm 3\sqrt{5}$

$m = -7 \pm 3\sqrt{5}$

46. $2x^2 - 2x - 1 = 0$

$2x^2 - 2x = 1$

$x^2 - x = \frac{1}{2}$

$x^2 - x + \frac{1}{4} = \frac{1}{2} + \frac{1}{4}$

$(x - \frac{1}{2})^2 = \frac{3}{4}$

$x - \frac{1}{2} = \pm\frac{\sqrt{3}}{2}$

$x = \frac{1}{2} \pm \frac{\sqrt{3}}{2}$

48. $4y^2 + 4y - 3 = 0$

$4y^2 + 4y = 3$

$y^2 + y = \frac{3}{4}$

$y^2 + y + \frac{1}{4} = \frac{3}{4} + \frac{1}{4}$

$(y + \frac{1}{2})^2 = 1$

$y + \frac{1}{2} = \pm1$

$y = -\frac{1}{2} \pm 1$

$y = \frac{1}{2}$ or $y = -\frac{3}{2}$

50. $2t^2 - t - 2 = 0$

$2t^2 - t = 2$

$t^2 - \frac{1}{2}t = 1$

$t^2 - \frac{1}{2}t + \frac{1}{16} = 1 + \frac{1}{16}$

$(t - \frac{1}{4})^2 = \frac{17}{16}$

$t - \frac{1}{4} = \pm\frac{\sqrt{17}}{4}$

$t = \frac{1}{4} \pm \frac{\sqrt{17}}{4}$

52. $3r^2 + 6r = 1$

$r^2 + 2r = \frac{1}{3}$

$r^2 + 2r + 1 = \frac{1}{3} + 1$

$(r + 1)^2 = \frac{4}{3}$

$r + 1 = \pm\frac{2}{\sqrt{3}}$

$r = -1 \pm \frac{2\sqrt{3}}{3}$

54. $3p^2 - 4p = 4$

$p^2 - \frac{4}{3}p = \frac{4}{3}$

$p^2 - \frac{4}{3}p + \frac{4}{9} = \frac{4}{3} + \frac{4}{9}$

$(p - \frac{2}{3})^2 = \frac{16}{9}$

$p - \frac{2}{3} = \pm\frac{4}{3}$

$p = \frac{2}{3} \pm \frac{4}{3}$

$p = 2$ or $p = -\frac{2}{3}$

56. $2x + x^2 = 2$

$x^2 + 2x = 2$

$x^2 + 2x + 1 = 2 + 1$

$(x + 1)^2 = 3$

$x + 1 = \pm\sqrt{3}$

$x = -1 \pm\sqrt{3}$

Since x is positive,

$x = -1 + \sqrt{3}.$

58. $x^2 - 4ax + 4a^2 - 9 = 0$

$\qquad x^2 - 4ax + 4a^2 = 9$

$\qquad (x - 2a)^2 = 9$

$\qquad x - 2a = \pm 3$

$\qquad x = 2a \pm 3$

60. $x = \dfrac{-1 + \sqrt{3}}{2} \approx 0.366$ $\qquad\qquad x = \dfrac{-1 - \sqrt{3}}{2} \approx -1.366$

$\qquad 2x^2 + 2x - 1 = 0 \qquad\qquad\qquad 2x^2 + 2x - 1 = 0$

$\qquad 2(0.366)^2 + 2(0.366) - 1 \overset{?}{=} 0 \qquad 2(-1.366)^2 + 2(-1.366) - 1 \overset{?}{=} 0$

$\qquad 0.267912 + 0.732 - 1 \overset{?}{=} 0 \qquad 3.731912 - 2.732 - 1 \overset{?}{=} 0$

$\qquad\qquad\qquad -0.000088 \approx 0 \qquad\qquad\qquad -0.000088 \approx 0$

Problem Set 6.3, pp. 250-251

2. a) $x^2 + x - 6 = 0$

$\qquad x = \dfrac{-1 \pm \sqrt{1^2 - 4(1)(-6)}}{2(1)}$

$\qquad x = \dfrac{-1 \pm \sqrt{1 + 24}}{2}$

$\qquad x = \dfrac{-1 \pm \sqrt{25}}{2}$

$\qquad x = \dfrac{-1 + 5}{2}$ or $x = \dfrac{-1 - 5}{2}$

$\qquad x = 2 \qquad\qquad x = -3$

b) $x^2 + x - 6 = 0$

$\qquad (x + 3)(x - 2) = 0$

$\qquad x + 3 = 0$ or $x - 2 = 0$

$\qquad\qquad x = -3 \qquad\qquad x = 2$

4. a) $5x^2 + 4x = 1$

$\qquad 5x^2 + 4x - 1 = 0$

$\qquad x = \dfrac{-4 \pm \sqrt{4^2 - 4(5)(-1)}}{2(5)}$

$\qquad x = \dfrac{-4 \pm \sqrt{16 + 20}}{10}$

b) $5x^2 + 4x = 1$

$\qquad 5x^2 + 4x - 1 = 0$

$\qquad (5x - 1)(x + 1) = 0$

$\qquad 5x - 1 = 0$ or $x + 1 = 0$

$\qquad\qquad x = \dfrac{1}{5} \qquad\qquad x = -1$

$$x = \frac{-4 \pm \sqrt{36}}{10}$$

$$x = \frac{-4 + 6}{10} \text{ or } x = \frac{-4 - 6}{10}$$

$$x = \frac{1}{5} \qquad\qquad x = -1$$

6. a) $x^2 - 10x + 25 = 0$

$$x = \frac{-(-10) \pm \sqrt{(-10)^2 - 4(1)(25)}}{2(1)}$$

$$x = \frac{10 \pm \sqrt{100 - 100}}{2}$$

$$x = \frac{10 \pm \sqrt{0}}{2}$$

$$x = \frac{10}{2} = 5$$

b) $x^2 - 10x + 25 = 0$

$$(x - 5)^2 = 0$$

$$x - 5 = 0$$

$$x = 5$$

8. a) $x^2 - 3x = 0$

$$x = \frac{-(-3) \pm \sqrt{(-3)^2 - 4(1)(0)}}{2(1)}$$

$$x = \frac{3 \pm \sqrt{9 - 0}}{2}$$

$$x = \frac{3 \pm \sqrt{9}}{2}$$

$$x = \frac{3 + 3}{2} \text{ or } x = \frac{3 - 3}{2}$$

$$x = 3 \qquad\qquad x = 0$$

b) $x^2 - 3x = 0$

$$x(x - 3) = 0$$

$$x = 0 \text{ or } x - 3 = 0$$

$$x = 3$$

10. a) $4x^2 - 1 = 0$

$$x = \frac{-0 \pm \sqrt{0^2 - 4(4)(-1)}}{2(4)}$$

$$x = \frac{0 \pm \sqrt{16}}{8}$$

$$x = \frac{\pm 4}{8}$$

$$x = \frac{1}{2} \text{ or } x = -\frac{1}{2}$$

b) $4x^2 - 1 = 0$

$$(2x - 1)(2x + 1) = 0$$

$$2x - 1 = 0 \text{ or } 2x + 1 = 0$$

$$x = \frac{1}{2} \qquad\qquad x = -\frac{1}{2}$$

12. $x^2 + x - 4 = 0$

$$x = \frac{-1 \pm \sqrt{1^2 - 4(1)(-4)}}{2(1)}$$

$$x = \frac{-1 \pm \sqrt{1 + 16}}{2}$$

$$x = \frac{-1 \pm \sqrt{17}}{2}$$

14. $y^2 - 5y - 2 = 0$

$$y = \frac{-(-5) \pm \sqrt{(-5)^2 - 4(1)(-2)}}{2(1)}$$

$$y = \frac{5 \pm \sqrt{25 + 8}}{2}$$

$$y = \frac{5 \pm \sqrt{33}}{2}$$

16. $p^2 + 3p - 9 = 0$

$$p = \frac{-3 \pm \sqrt{3^2 - 4(1)(-9)}}{2(1)}$$

$$p = \frac{-3 \pm \sqrt{9 + 36}}{2}$$

$$p = \frac{-3 \pm \sqrt{45}}{2}$$

$$p = \frac{-3 \pm 3\sqrt{5}}{2}$$

18. $r^2 = 4r + 7$

$r^2 - 4r - 7 = 0$

$$r = \frac{-(-4) \pm \sqrt{(-4)^2 - 4(1)(-7)}}{2(1)}$$

$$r = \frac{4 \pm \sqrt{16 + 28}}{2}$$

$$r = \frac{4 \pm \sqrt{44}}{2}$$

$$r = \frac{4 \pm 2\sqrt{11}}{2}$$

$$r = 2 \pm \sqrt{11}$$

20. $2x^2 + 20x + 36 = 0$

$x^2 + 10x + 18 = 0$

$$x = \frac{-10 \pm \sqrt{10^2 - 4(1)(18)}}{2(1)}$$

$$x = \frac{-10 \pm \sqrt{100 - 72}}{2}$$

$$x = \frac{-10 \pm \sqrt{28}}{2}$$

$$x = \frac{-10 \pm 2\sqrt{7}}{2}$$

$$x = -5 \pm \sqrt{7}$$

22. $2m^2 - m - 4 = 0$

$$m = \frac{-(-1) \pm \sqrt{(-1)^2 - 4(2)(-4)}}{2(2)}$$

$$m = \frac{1 \pm \sqrt{1 + 32}}{4}$$

$$m = \frac{1 \pm \sqrt{33}}{4}$$

24. $2z^2 - 4z = 3$

$2z^2 - 4z - 3 = 0$

$z = \dfrac{-(-4) \pm \sqrt{(-4)^2 - 4(2)(-3)}}{2(2)}$

$z = \dfrac{4 \pm \sqrt{16 + 24}}{4}$

$z = \dfrac{4 \pm \sqrt{40}}{4}$

$z = \dfrac{4 \pm 2\sqrt{10}}{4}$

$z = \dfrac{2(2 \pm \sqrt{10})}{4}$

$z = \dfrac{2 \pm \sqrt{10}}{2}$

26. $3t^2 - 2t - 4 = 0$

$t = \dfrac{-(-2) \pm \sqrt{(-2)^2 - 4(3)(-4)}}{2(3)}$

$t = \dfrac{2 \pm \sqrt{4 + 48}}{6}$

$t = \dfrac{2 \pm \sqrt{52}}{6}$

$t = \dfrac{2 \pm 2\sqrt{13}}{6}$

$t = \dfrac{2(1 \pm \sqrt{13})}{6}$

$t = \dfrac{1 \pm \sqrt{13}}{3}$

28. $(2p + 3)(2p + 1) = 2$

$4p^2 + 8p + 3 = 2$

$4p^2 + 8p + 1 = 0$

$p = \dfrac{-8 \pm \sqrt{8^2 - 4(4)(1)}}{2(4)}$

$p = \dfrac{-8 \pm \sqrt{64 - 16}}{8}$

$p = \dfrac{-8 \pm \sqrt{48}}{8}$

$p = \dfrac{-8 \pm 4\sqrt{3}}{8}$

$p = \dfrac{4(-2 \pm \sqrt{3})}{8}$

$p = \dfrac{-2 \pm \sqrt{3}}{2}$

30. $5r(r + 2) = -4$

$5r^2 + 10r = -4$

$5r^2 + 10r + 4 = 0$

$r = \dfrac{-10 \pm \sqrt{10^2 - 4(5)(4)}}{2(5)}$

$r = \dfrac{-10 \pm \sqrt{100 - 80}}{10}$

$r = \dfrac{-10 \pm \sqrt{20}}{10}$

$r = \dfrac{-10 \pm 2\sqrt{5}}{10}$

$r = \dfrac{2(-5 \pm \sqrt{5})}{10}$

$r = \dfrac{-5 \pm \sqrt{5}}{5}$

32.

$$90m^2 = 20 - 60m$$

$$90m^2 + 60m - 20 = 0$$

$$9m^2 + 6m - 2 = 0$$

$$m = \frac{-6 \pm \sqrt{6^2 - 4(9)(-2)}}{2(9)}$$

$$m = \frac{-6 \pm \sqrt{36 + 72}}{18}$$

$$m = \frac{-6 \pm \sqrt{108}}{18}$$

$$m = \frac{-6 \pm 6\sqrt{3}}{18}$$

$$m = \frac{6(-1 \pm \sqrt{3})}{18}$$

$$m = \frac{-1 \pm \sqrt{3}}{3}$$

34.

$$\frac{y^2}{4} + \frac{y}{3} - \frac{5}{12} = 0$$

$$12\left(\frac{y^2}{4} + \frac{y}{3} - \frac{5}{12}\right) = 12 \cdot 0$$

$$3y^2 + 4y - 5 = 0$$

$$y = \frac{-4 \pm \sqrt{4^2 - 4(3)(-5)}}{2(3)}$$

$$y = \frac{-4 \pm \sqrt{16 + 60}}{6}$$

$$y = \frac{-4 \pm \sqrt{76}}{6}$$

$$y = \frac{-4 \pm 2\sqrt{19}}{6}$$

$$y = \frac{2(-2 \pm \sqrt{19})}{6}$$

$$y = \frac{-2 \pm \sqrt{19}}{3}$$

36.

$$\frac{2}{3}x^2 + \frac{4}{9}x - \frac{1}{3} = 0$$

$$9\left(\frac{2}{3}x^2 + \frac{4}{9}x - \frac{1}{3}\right) = 9 \cdot 0$$

$$6x^2 + 4x - 3 = 0$$

$$x = \frac{-4 \pm \sqrt{4^2 - 4(6)(-3)}}{2(6)}$$

$$x = \frac{-4 \pm \sqrt{16 + 72}}{12}$$

$$x = \frac{-4 \pm \sqrt{88}}{12}$$

$$x = \frac{-4 \pm 2\sqrt{22}}{12}$$

$$x = \frac{2(-2 \pm \sqrt{22})}{12}$$

$$x = \frac{-2 \pm \sqrt{22}}{6}$$

38.

$$0.3z^2 - 1.4z - 0.4 = 0$$

$$10(0.3z^2 - 1.4z - 0.4) = 10 \cdot 0$$

$$3z^2 - 14z - 4 = 0$$

$$z = \frac{-(-14) \pm \sqrt{(-14)^2 - 4(3)(-4)}}{2(3)}$$

$$z = \frac{14 \pm \sqrt{196 + 48}}{6}$$

$$z = \frac{14 \pm \sqrt{244}}{6}$$

$$z = \frac{14 \pm 2\sqrt{61}}{6}$$

$$z = \frac{2(7 \pm \sqrt{61})}{6}$$

$$z = \frac{7 \pm \sqrt{61}}{3}$$

40.
$$2x^2 - \sqrt{3}x = \frac{3}{4}$$
$$4(2x^2 - \sqrt{3}x) = 4\left(\frac{3}{4}\right)$$
$$8x^2 - 4\sqrt{3}x = 3$$
$$8x^2 - 4\sqrt{3}x - 3 = 0$$
$$x = \frac{-(-4\sqrt{3}) \pm \sqrt{(-4\sqrt{3})^2 - 4(8)(-3)}}{2(8)}$$
$$x = \frac{4\sqrt{3} \pm \sqrt{16 \cdot 3 + 96}}{16}$$
$$x = \frac{4\sqrt{3} \pm \sqrt{144}}{16}$$
$$x = \frac{4\sqrt{3} \pm 12}{16} = \frac{4(\sqrt{3} \pm 3)}{16} = \frac{\sqrt{3} \pm 3}{4}$$

42.
$$x^2 = 4x - 2$$
$$x^2 - 4x + 2 = 0$$
$$x = \frac{-(-4) \pm \sqrt{(-4)^2 - 4(1)(2)}}{2(1)}$$
$$x = \frac{4 \pm \sqrt{16 - 8}}{2}$$
$$x = \frac{4 \pm \sqrt{8}}{2}$$
$$x = \frac{4 \pm 2\sqrt{2}}{2}$$
$$x = 2 \pm \sqrt{2}$$

44.
$$76 = -16t^2 + 80t + 20$$
$$16t^2 - 80t + 56 = 0$$
$$4t^2 - 20t + 14 = 0$$
$$t = \frac{-(-20) \pm \sqrt{(-20)^2 - 4(4)(14)}}{2(4)}$$
$$t = \frac{20 \pm \sqrt{400 - 224}}{8}$$

$$t = \frac{20 \pm \sqrt{176}}{8}$$

$$t = \frac{20 \pm 4\sqrt{11}}{8} = \frac{4(5 \pm \sqrt{11})}{8} = \frac{5 \pm \sqrt{11}}{2}$$

$$t = \frac{5 - \sqrt{11}}{2} \approx 0.84 \text{ sec and } t = \frac{5 + \sqrt{11}}{2} \approx 4.16 \text{ sec}$$

46.   R = C

$$175x - \frac{1}{4}x^2 = 20x$$

$$-\frac{1}{4}x^2 + 155x = 0$$

$$4(-\frac{1}{4}x^2 + 155x) = 4 \cdot 0$$

$$-x^2 + 620x = 0$$

$$-x(x - 620) = 0$$

$$-x = 0 \text{ or } x - 620 = 0$$

$$x = 0 \qquad\qquad x = 620$$

48. $1.47x^2 - 7.04x + 6.32 = 0$

$$x = \frac{-(-7.04) \pm \sqrt{(-7.04)^2 - 4(1.47)(6.32)}}{2(1.47)}$$

$$x = \frac{7.04 \pm \sqrt{49.5616 - 37.1616}}{2.94}$$

$$x = \frac{7.04 \pm \sqrt{12.4}}{2.94}$$

$$x \approx \frac{7.04 \pm 3.521}{2.94}$$

$$x \approx \frac{7.04 + 3.521}{2.94} \quad \text{or} \quad x \approx \frac{7.04 - 3.521}{2.94}$$

$$x \approx 3.59 \qquad\qquad x \approx 1.20$$

Problem Set 6.4, pp. 255-257

2. x = first number

14 − x = second number

Product = 45

$x(14 - x) = 45$

$14x - x^2 = 45$

$0 = x^2 - 14x + 45$

$0 = (x - 5)(x - 9)$

x = 5 or x = 9

14 − x = 9   14 − x = 5

4. x = first number

2x − 5 = second number

Product = 7

$x(2x - 5) = 7$

$2x^2 - 5x = 7$

$2x^2 - 5x - 7 = 0$

$(2x - 7)(x + 1) = 0$

$x = \frac{7}{2}$ or x ✕ −1

2x − 5 = 2

Eliminate x = −1 since it is negative.

6. x = first integer

x + 1 = second integer

Sum of squares = 113

$x^2 + (x + 1)^2 = 113$

$x^2 + x^2 + 2x + 1 = 113$

$2x^2 + 2x - 112 = 0$

$x^2 + x - 56 = 0$

$(x + 8)(x - 7) = 0$

x ✕ −8 or x = 7

x + 1 = 8

Eliminate x = −8 since it is negative.

8. x = first integer

x + 2 = second integer

Square of sum = 64

$(x + x + 2)^2 = 64$

$(2x + 2)^2 = 64$

$4x^2 + 8x + 4 = 64$

$4x^2 + 8x - 60 = 0$

$x^2 + 2x - 15 = 0$

$(x + 5)(x - 3) = 0$

x = −5 or x = 3

x + 2 = −3   x + 2 = 5

10.   x = first integer

x + 2 = second integer

x + 4 = third integer

Square of first plus product of other two = 224

$$x^2 + (x + 2)(x + 4) = 224$$

$$x^2 + x^2 + 6x + 8 = 224$$

$$2x^2 + 6x - 216 = 0$$

$$x^2 + 3x - 108 = 0$$

$$(x + 12)(x - 9) = 0$$

$$x = -12 \quad \text{or} \quad x = 9$$

$$x + 2 = -10$$

$$x + 4 = -8$$

Eliminate x = 9 since it is odd.

12.   x = width of rectangle

3x + 2 = length of rectangle

Area = 208

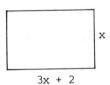

$$x(3x + 2) = 208$$

$$3x^2 + 2x = 208$$

$$3x^2 + 2x - 208 = 0$$

$$x = \frac{-2 \pm \sqrt{2^2 - 4(3)(-208)}}{2(3)}$$

$$x = \frac{-2 \pm \sqrt{4 + 2496}}{6}$$

$$x = \frac{-2 \pm \sqrt{2500}}{6}$$

$$x = \frac{-2 + 50}{6} \quad \text{or} \quad x = \frac{-2 - 50}{6}$$

$$x = 8 \text{ m} \qquad \cancel{x = -\frac{26}{3}}$$

3x + 2 = 26 m

14. x = width of border

$$\begin{matrix}\text{Area of 4}\\\text{corners}\end{matrix} + \begin{matrix}\text{area of}\\\text{2 sides}\end{matrix} + \begin{matrix}\text{area of top}\\\text{and bottom}\end{matrix} = 336$$

$$4x^2 \quad + \quad 2(20x) \quad + \quad 2(30x) \quad = 336$$

$$4x^2 + 40x + 60x = 336$$

$$4x^2 + 100x - 336 = 0$$

$$x^2 + 25x - 84 = 0$$

$$(x + 28)(x - 3) = 0$$

$$x = -28 \quad \text{or} \quad x = 3 \text{ ft}$$

16.   Volume = 720

$$\ell wh = 720$$

$$(x - 10)(x - 10)5 = 720$$

$$(x - 10)^2 = 144$$

$$x - 10 = \pm 12$$

x - 10 = 12   or   x - 10 = -12

x = 22 in. $x = -2$

18.   x = shorter leg

2x + 2 = longer leg

3x - 2 = hypotenuse

$$c^2 = a^2 + b^2$$

$$(3x - 2)^2 = x^2 + (2x + 2)^2$$

$$9x^2 - 12x + 4 = x^2 + 4x^2 + 8x +$$

$$4x^2 - 20x = 0$$

$$4x(x - 5) = 0$$

4x = 0   or   x - 5 = 0

$x = 0$ x = 5 in.

2x + 2 = 12 in.

3x - 2 = 13 in.

20. x = length of a side

$$a^2 + b^2 = c^2$$

$$x^2 + x^2 = 6^2$$

$$2x^2 = 36$$

$$x^2 = 18$$

$$x = \sqrt{18} = 3\sqrt{2} \text{ ft}$$

22.   x = rate of A

    x + 10 = rate of B

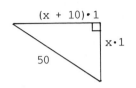

$$a^2 + b^2 = c^2$$

$$x^2 + (x + 10)^2 = 50^2$$

$$x^2 + x^2 + 20x + 100 = 2500$$

$$2x^2 + 20x - 2400 = 0$$

$$x^2 + 10x - 1200 = 0$$

$$(x + 40)(x - 30) = 0$$

x ~~= -40~~ or x = 30 mph

                x + 10 = 40 mph

24. $x^2 + 3^2 = 6^2$

    $x^2 + 9 = 36$

     $x^2 = 27$

      $x = \sqrt{27} = 3\sqrt{3}$

Height = $3 + 3\sqrt{3} + 3 = 6 + 3\sqrt{3}$ in.

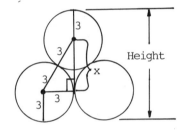

26. $1151^2 \stackrel{?}{=} 577^2 + 997^2$

   $1{,}324{,}801 \stackrel{?}{=} 332{,}929 + 994{,}009$

   $1{,}324{,}801 \neq 1{,}326{,}938$

Problem Set 6.5, pp. 261–263

2. $\sqrt{x} = 9$

    $(\sqrt{x})^2 = 9^2$

      $x = 81$

4. $\sqrt{x - 2} = 3$

    $(\sqrt{x - 2})^2 = 3^2$

     $x - 2 = 9$

      $x = 11$

6. $\sqrt{2x - 5} = 11$

$(\sqrt{2x - 5})^2 = 11^2$

$2x - 5 = 121$

$2x = 126$

$x = 63$

8. $\sqrt{x} - 3 = 5$

$\sqrt{x} = 8$

$(\sqrt{x})^2 = 8^2$

$x = 64$

10. $\sqrt{3y + 1} - 10 = 0$

$\sqrt{3y + 1} = 10$

$(\sqrt{3y + 1})^2 = 10^2$

$3y + 1 = 100$

$3y = 99$

$y = 33$

12. $\sqrt{4z - 7} + 1 = 0$

$\sqrt{4z - 7} = -1$

$(\sqrt{4z - 7})^2 = (-1)^2$

$4z - 7 = 1$

$4z = 8$

$z = 2$

Since 2 does not check, there is no solution.

14. $\sqrt{5p + 9} + 2 = 5$

$\sqrt{5p + 9} = 3$

$(\sqrt{5p + 9})^2 = 3^2$

$5p + 9 = 9$

$5p = 0$

$p = 0$

16. $\sqrt{v - 7} = \sqrt{7 - v}$

$(\sqrt{v - 7})^2 = (\sqrt{7 - v})^2$

$v - 7 = 7 - v$

$2v = 14$

$v = 7$

18. $\sqrt{6r} - 2\sqrt{r + 5} = 0$

$\sqrt{6r} = 2\sqrt{r + 5}$

$(\sqrt{6r})^2 = (2\sqrt{r + 5})^2$

$6r = 4(r + 5)$

$6r = 4r + 20$

$2r = 20$

$r = 10$

20. $\sqrt{s + 3} = \sqrt{s + 4}$

$(\sqrt{s + 3})^2 = (\sqrt{s + 4})^2$

$s + 3 = s + 4$

$3 = 4$

No solution

22.
$$\sqrt{4 + 9\sqrt{x}} = 7$$
$$(\sqrt{4 + 9\sqrt{x}})^2 = 7^2$$
$$4 + 9\sqrt{x} = 49$$
$$9\sqrt{x} = 45$$
$$\sqrt{x} = 5$$
$$(\sqrt{x})^2 = 5^2$$
$$x = 25$$

24.
$$\sqrt{m^2 + 45} = m - 3$$
$$(\sqrt{m^2 + 45})^2 = (m - 3)^2$$
$$m^2 + 45 = m^2 - 6m + 9$$
$$36 = -6m$$
$$-6 = m$$

Since -6 does not check, there is no solution.

26.
$$\sqrt{r^2 - 7r + 21} = r$$
$$(\sqrt{r^2 - 7r + 21})^2 = r^2$$
$$r^2 - 7r + 21 = r^2$$
$$-7r + 21 = 0$$
$$-7r = -21$$
$$r = 3$$

28.
$$\sqrt{x - 5} = \sqrt{x} - 1$$
$$(\sqrt{x - 5})^2 = (\sqrt{x} - 1)^2$$
$$x - 5 = x - 2\sqrt{x} + 1$$
$$-6 = -2\sqrt{x}$$
$$3 = \sqrt{x}$$
$$3^2 = (\sqrt{x})^2$$
$$9 = x$$

30.
$$\sqrt{y + 3} + 1 = \sqrt{y + 5}$$
$$(\sqrt{y + 3} + 1)^2 = (\sqrt{y + 5})^2$$
$$y + 3 + 2\sqrt{y + 3} + 1 = y + 5$$
$$2\sqrt{y + 3} = 1$$
$$(2\sqrt{y + 3})^2 = 1^2$$
$$4(y + 3) = 1$$
$$4y + 12 = 1$$
$$4y = -11$$
$$y = -\frac{11}{4}$$

32. $\sqrt{x^2 + 5} = 3$

$(\sqrt{x^2 + 5})^2 = 3^2$

$x^2 + 5 = 9$

$x^2 = 4$

$x = \pm 2$

34. $\sqrt{x^2 - 1} = 2$

$(\sqrt{x^2 - 1})^2 = 2^2$

$x^2 - 1 = 4$

$x^2 = 5$

$x = \pm\sqrt{5}$

36. $\sqrt{x + 6} = x$

$(\sqrt{x + 6})^2 = x^2$

$x + 6 = x^2$

$0 = x^2 - x - 6$

$0 = (x - 3)(x + 2)$

$x = 3$ or $x = -2$

Since -2 does not check, the only solution is 3.

38. $\sqrt{10y - 1} - 3y = 0$

$\sqrt{10y - 1} = 3y$

$(\sqrt{10y - 1})^2 = (3y)^2$

$10y - 1 = 9y^2$

$0 = 9y^2 - 10y + 1$

$0 = (9y - 1)(y - 1)$

$y = \frac{1}{9}$ or $y = 1$

40. $\sqrt{p} = p - 2$

$(\sqrt{p})^2 = (p - 2)^2$

$p = p^2 - 4p + 4$

$0 = p^2 - 5p + 4$

$0 = (p - 1)(p - 4)$

$p = 1$ or $p = 4$

Since 1 does not check, the only solution is 4.

42. $\sqrt{r^2 + 12r} = 2\sqrt{7}$

$(\sqrt{r^2 + 12r})^2 = (2\sqrt{7})^2$

$r^2 + 12r = 4(7)$

$r^2 + 12r - 28 = 0$

$(r + 14)(r - 2) = 0$

$r = -14$ or $r = 2$

44. $\sqrt{m + 9} = m - 3$

$(\sqrt{m + 9})^2 = (m - 3)^2$

$m + 9 = m^2 - 6m + 9$

$0 = m^2 - 7m$

$0 = m(m - 7)$

$m = 0$ or $m = 7$

Since 0 does not check, the only solution is 7.

46. $\sqrt{t - 2} + t - 4 = 0$

$\sqrt{t - 2} = 4 - t$

$(\sqrt{t - 2})^2 = (4 - t)^2$

$t - 2 = 16 - 8t + t^2$

$0 = t^2 - 9t + 18$

$0 = (t - 3)(t - 6)$

$t = 3$ or $t = 6$

Since 6 does not check, the only solution is 3.

48. $\sqrt{2x^2 - 11x + 13} = 3 - x$

$(\sqrt{2x^2 - 11x + 13})^2 = (3 - x)^2$

$2x^2 - 11x + 13 = 9 - 6x + x^2$

$x^2 - 5x + 4 = 0$

$(x - 1)(x - 4) = 0$

$x = 1$ or $x = 4$

Since 4 does not check, the only solution is 1.

50. $\sqrt{3y + 4} - \sqrt{y} = 2$

$\sqrt{3y + 4} = 2 + \sqrt{y}$

$(\sqrt{3y + 4})^2 = (2 + \sqrt{y})^2$

$3y + 4 = 4 + 4\sqrt{y} + y$

$2y = 4\sqrt{y}$

$y = 2\sqrt{y}$

$y^2 = (2\sqrt{y})^2$

$y^2 = 4y$

$y^2 - 4y = 0$

$y(y - 4) = 0$

$y = 0$ or $y = 4$

52. $\sqrt{5p + 6} - \sqrt{3p + 4} = 2$

$\sqrt{5p + 6} = 2 + \sqrt{3p + 4}$

$(\sqrt{5p + 6})^2 = (2 + \sqrt{3p + 4})^2$

$5p + 6 = 4 + 4\sqrt{3p + 4} + (3p + 4)$

$2p - 2 = 4\sqrt{3p + 4}$

$p - 1 = 2\sqrt{3p + 4}$

$(p - 1)^2 = (2\sqrt{3p + 4})^2$

$p^2 - 2p + 1 = 4(3p + 4)$

$p^2 - 2p + 1 = 12p + 16$

$$p^2 - 14p - 15 = 0$$

$$(p + 1)(p - 15) = 0$$

$$p = -1 \text{ or } p = 15$$

Since -1 does not check, the only solution is 15.

54. $\sqrt{5r + 6} + \sqrt{r + 3} = 3$

$$\sqrt{5r + 6} = 3 - \sqrt{r + 3}$$

$$(\sqrt{5r + 6})^2 = (3 - \sqrt{r + 3})^2$$

$$5r + 6 = 9 - 6\sqrt{r + 3} + (r + 3)$$

$$4r - 6 = -6\sqrt{r + 3}$$

$$2r - 3 = -3\sqrt{r + 3}$$

$$(2r - 3)^2 = (-3\sqrt{r + 3})^2$$

$$4r^2 - 12r + 9 = 9(r + 3)$$

$$4r^2 - 12r + 9 = 9r + 27$$

$$4r^2 - 21r - 18 = 0$$

$$(4r + 3)(r - 6) = 0$$

$$r = -\frac{3}{4} \text{ or } r = 6$$

Since 6 does not check, the only solution is $-\frac{3}{4}$.

56. $\sqrt[3]{x} = 5$

$(\sqrt[3]{x})^3 = 5^3$

$x = 125$

58. $\sqrt[4]{x - 1} = 2$

$(\sqrt[4]{x - 1})^4 = 2^4$

$x - 1 = 16$

$x = 17$

60. $\sqrt[3]{x + 5} + 1 = 0$

$\sqrt[3]{x + 5} = -1$

$(\sqrt[3]{x + 5})^3 = (-1)^3$

$x + 5 = -1$

$x = -6$

62. $\sqrt[3]{3t - 2} = \sqrt[3]{5t + 2}$

$(\sqrt[3]{3t - 2})^3 = (\sqrt[3]{5t + 2})^3$

$3t - 2 = 5t + 2$

$-4 = 2t$

$-2 = t$

64.
$$\sqrt[3]{y^2 + 23} = 3$$
$$(\sqrt[3]{y^2 + 23})^3 = 3^3$$
$$y^2 + 23 = 27$$
$$y^2 = 4$$
$$y = \pm 2$$

66.
$$\sqrt[4]{\sqrt{r} - 44} = 2$$
$$(\sqrt[4]{\sqrt{r} - 44})^4 = 2^4$$
$$\sqrt{r} - 44 = 16$$
$$\sqrt{r} = 60$$
$$(\sqrt{r})^2 = 60^2$$
$$r = 3600$$

68.
$$v = \sqrt{2gR}$$
$$v^2 = (\sqrt{2gR})^2$$
$$v^2 = 2gR$$
$$\frac{v^2}{2g} = R$$

70.
$$r = \sqrt{\frac{S}{4\pi}}$$
$$r^2 = (\sqrt{\frac{S}{4\pi}})^2$$
$$r^2 = \frac{S}{4\pi}$$
$$4\pi r^2 = S$$

72.
$$r = \sqrt[3]{\frac{3V}{4\pi}}$$
$$r^3 = (\sqrt[3]{\frac{3V}{4\pi}})^3$$
$$r^3 = \frac{3V}{4\pi}$$
$$4\pi r^3 = 3V$$
$$\frac{4\pi r^3}{3} = V$$

74.
$$c = \sqrt{a^2 + b^2}$$
$$c^2 = (\sqrt{a^2 + b^2})^2$$
$$c^2 = a^2 + b^2$$
$$c^2 - a^2 = b^2$$
$$\pm\sqrt{c^2 - a^2} = b$$

76.
$$\sqrt{x + 7} = x - 5$$
$$(\sqrt{x + 7})^2 = (x - 5)^2$$
$$x + 7 = x^2 - 10x + 25$$
$$0 = x^2 - 11x + 18$$
$$0 = (x - 2)(x - 9)$$
$$x = 2 \text{ or } x = 9$$

Since 2 does not check, the only solution is 9.

78.
$$t = 2\pi\sqrt{\frac{\ell}{32}}$$
$$6 = 2\pi\sqrt{\frac{\ell}{32}}$$
$$\frac{6}{2\pi} = \sqrt{\frac{\ell}{32}}$$
$$\frac{3}{\pi} = \sqrt{\frac{\ell}{32}}$$
$$(\frac{3}{\pi})^2 = (\sqrt{\frac{\ell}{32}})^2$$
$$\frac{9}{\pi^2} = \frac{\ell}{32}$$
$$\frac{288}{\pi^2} \text{ ft} = \ell$$

80. $d = \sqrt{1.5a}$

$178 = \sqrt{1.5a}$

$178^2 = (\sqrt{1.5a})^2$

$31{,}684 = 1.5a$

$a = \dfrac{31{,}684}{1.5} \approx 21{,}123 \text{ ft} = \dfrac{21{,}123}{5280} \text{ mi} \approx 4 \text{ mi}$

Problem Set 6.6, pp. 267-268

2. $x - \dfrac{4}{x} = 0$

$x(x - \dfrac{4}{x}) = x \cdot 0$

$x^2 - 4 = 0$

$x^2 = 4$

$x = \pm 2$

4. $1 + \dfrac{8}{r} + \dfrac{12}{r^2} = 0$

$r^2(1 + \dfrac{8}{r} + \dfrac{12}{r^2}) = r^2 \cdot 0$

$r^2 + 8r + 12 = 0$

$(r + 2)(r + 6) = 0$

$r = -2 \text{ or } r = -6$

6. $6y + 13 = \dfrac{15}{y}$

$y(6y + 13) = y(\dfrac{15}{y})$

$6y^2 + 13y = 15$

$6y^2 + 13y - 15 = 0$

$(6y - 5)(y + 3) = 0$

$y = \dfrac{5}{6} \text{ or } y = -3$

8. $1 + \dfrac{5}{6p} = \dfrac{1}{p^2}$

$6p^2(1 + \dfrac{5}{6p}) = 6p^2(\dfrac{1}{p^2})$

$6p^2 + 5p = 6$

$6p^2 + 5p - 6 = 0$

$(3p - 2)(2p + 3) = 0$

$p = \dfrac{2}{3} \text{ or } p = -\dfrac{3}{2}$

10. $\dfrac{t}{t - 5} + \dfrac{3}{t - 5} + t = 0$

$(t - 5)\dfrac{t}{t - 5} + (t - 5)\dfrac{3}{t - 5} + (t - 5)t = (t - 5) \cdot 0$

$t + 3 + (t - 5)t = 0$

$t + 3 + t^2 - 5t = 0$

$t^2 - 4t + 3 = 0$

$(t - 1)(t - 3) = 0$

$t = 1 \text{ or } t = 3$

12.
$$\frac{1}{x} + \frac{1}{x - 1} = \frac{7}{12}$$

$$12x(x - 1)\frac{1}{x} + 12x(x - 1)\frac{1}{x - 1} = 12x(x - 1)\frac{7}{12}$$

$$12(x - 1) + 12x = x(x - 1)7$$

$$12x - 12 + 12x = 7x^2 - 7x$$

$$0 = 7x^2 - 31x + 12$$

$$0 = (7x - 3)(x - 4)$$

$$x = \frac{3}{7} \text{ or } x = 4$$

14.
$$\frac{m}{m + 2} + \frac{4}{m - 1} = 2$$

$$(m + 2)(m - 1)\frac{m}{m + 2} + (m + 2)(m - 1)\frac{4}{m - 1} = (m + 2)(m - 1)2$$

$$(m - 1)m + (m + 2)4 = (m^2 + m - 2)2$$

$$m^2 - m + 4m + 8 = 2m^2 + 2m - 4$$

$$0 = m^2 - m - 12$$

$$0 = (m - 4)(m + 3)$$

$$m = 4 \text{ or } m = -3$$

16.
$$1 = \frac{1}{y + 3} + \frac{20}{(y + 3)^2}$$

$$(y + 3)^2 1 = (y + 3)^2 \frac{1}{y + 3} + (y + 3)^2 \frac{20}{(y + 3)^2}$$

$$y^2 + 6y + 9 = (y + 3) + 20$$

$$y^2 + 5y - 14 = 0$$

$$(y + 7)(y - 2) = 0$$

$$y = -7 \text{ or } y = 2$$

18.
$$\frac{3}{r^2 - 3r} - \frac{1}{r - 3} = 1$$

$$\frac{3}{r(r - 3)} - \frac{1}{r - 3} = 1$$

$$r(r - 3)\frac{3}{r(r - 3)} - r(r - 3)\frac{1}{r - 3} = r(r - 3)1$$

$$3 - r = r^2 - 3r$$

$$0 = r^2 - 2r - 3$$

$$0 = (r - 3)(r + 1)$$

$$r = 3 \text{ or } r = -1$$

Since 3 makes both denominators 0, the only solution is -1.

20. $x^4 - 8x^2 + 16 = 0$

$(x^2 - 4)(x^2 - 4) = 0$

$x^2 - 4 = 0$

$x^2 = 4$

$x = \pm 2$

22. $x^4 - 13x^2 + 36 = 0$

$(x^2 - 4)(x^2 - 9) = 0$

$x^2 - 4 = 0$ or $x^2 - 9 = 0$

$x^2 = 4 \qquad x^2 = 9$

$x = \pm 2 \qquad x = \pm 3$

24. $t^4 - 7t^2 + 12 = 0$

$(t^2 - 4)(t^2 - 3) = 0$

$t^2 - 4 = 0$ or $t^2 - 3 = 0$

$t^2 = 4 \qquad t^2 = 3$

$t = \pm 2 \qquad t = \pm\sqrt{3}$

26. $4y^4 - 29y^2 + 25 = 0$

$(y^2 - 1)(4y^2 - 25) = 0$

$y^2 - 1 = 0$ or $4y^2 - 25 = 0$

$y^2 = 1 \qquad y^2 = \frac{25}{4}$

$y = \pm 1 \qquad y = \pm\frac{5}{2}$

28. $x^{1/2} - 3x^{1/4} + 2 = 0$

Let $u = x^{1/4}$. Then $u^2 = x^{1/2}$.

$u^2 - 3u + 2 = 0$

$(u - 1)(u - 2) = 0$

$u = 1$ or $\qquad u = 2$

$x^{1/4} = 1 \qquad x^{1/4} = 2$

$(x^{1/4})^4 = 1^4 \quad (x^{1/4})^4 = 2^4$

$x = 1 \qquad\qquad x = 16$

30. $p^{2/3} - p^{1/3} - 6 = 0$

Let $u = p^{1/3}$. Then $u^2 = p^{2/3}$

$u^2 - u - 6 = 0$

$(u - 3)(u + 2) = 0$

$u = 3$ or $\qquad u = -2$

$p^{1/3} = 3 \qquad p^{1/3} = -2$

$(p^{1/3})^3 = 3^3 \quad (p^{1/3})^3 = (-2)^3$

$p = 27 \qquad\qquad p = -8$

32. $7r^{-2} + 41r^{-1} - 6 = 0$

Let $u = r^{-1}$. Then $u^2 = r^{-2}$.

$7u^2 + 41u - 6 = 0$

$(7u - 1)(u + 6) = 0$

$$u = \frac{1}{7} \qquad \text{or} \qquad u = -6$$

$$r^{-1} = \frac{1}{7} \qquad\qquad r^{-1} = -6$$

$$(r^{-1})^{-1} = \left(\frac{1}{7}\right)^{-1} \qquad (r^{-1})^{-1} = (-6)^{-1}$$

$$r = 7 \qquad\qquad r = -\frac{1}{6}$$

34. $t = $ time for new press

$t + 8 = $ time for old press

Part of job done by new press in 1 hr	+	part of job done by old press in 1 hr	=	part of job done by both presses in 1 hr
$\frac{1}{t}$	+	$\frac{1}{t + 8}$	=	$\frac{1}{3}$

$$3t(t + 8)\frac{1}{t} + 3t(t + 8)\frac{1}{t + 8} = 3t(t + 8)\frac{1}{3}$$

$$3(t + 8) + 3t = t(t + 8)$$

$$3t + 24 + 3t = t^2 + 8t$$

$$0 = t^2 + 2t - 24$$

$$0 = (t + 6)(t - 4)$$

$$t = -6 \text{ or } \qquad t = 4 \text{ hr}$$

$$t + 8 = 12 \text{ hr}$$

36.   x = speed to the city

x + 20 = speed on return trip

$$\text{Time to the city} - \text{time for return trip} = \frac{1}{2}$$

$$\frac{60}{x} - \frac{60}{x + 20} = \frac{1}{2}$$

$$2x(x + 20)\frac{60}{x} - 2x(x + 20)\frac{60}{x + 20} = 2x(x + 20)\frac{1}{2}$$

$$2(x + 20)60 - 2x(60) = x(x + 20)$$

$$120x + 2400 - 120x = x^2 + 20x$$

$$0 = x^2 + 20x - 2400$$

$$0 = (x + 60)(x - 40)$$

$$x = -60 \text{ or } \qquad x = 40 \text{ mph}$$

$$x + 20 = 60 \text{ mph}$$

38.   x = speed of current

$$\text{Time upstream} - \text{time downstream} = 1$$

$$\frac{24}{10 - x} - \frac{24}{10 + x} = 1$$

$$(10 - x)(10 + x)\frac{24}{10 - x} - (10 - x)(10 + x)\frac{24}{10 + x} = (10 - x)(10 + x)1$$

$$(10 + x)24 - (10 - x)24 = 100 - x^2$$

$$240 + 24x - 240 + 24x = 100 - x^2$$

$$x^2 + 48x - 100 = 0$$

$$(x + 50)(x - 2) = 0$$

$$x = -50 \text{ or } x = 2 \text{ mph}$$

40. $I = \dfrac{W}{4\pi r^2}$

$4\pi r^2 I = W$

$r^2 = \dfrac{W}{4\pi I}$

$r = \pm\sqrt{\dfrac{W}{4\pi I}}$

42. x = speed of plane in still air

$$\frac{\text{Time against}}{\text{wind}} + \frac{\text{time with}}{\text{wind}} = 8\frac{1}{4} \text{ hr}$$

$$\frac{220}{x - 20} + \frac{220}{x + 20} = \frac{33}{4}$$

Multiply by $4(x - 20)(x + 20)$.

$$880(x + 20) + 880(x - 20) = (x^2 - 400)33$$

$$880x + 17,600 + 880x - 17,600 = 33x^2 - 13,200$$

$$-33x^2 + 1760x + 13,200 = 0$$

$$3x^2 - 160x - 1200 = 0 \qquad \text{Divide by } -11$$

$$x = \frac{-(-160) \pm \sqrt{(-160)^2 - 4(3)(-1200)}}{2(3)}$$

$$x = \frac{160 \pm \sqrt{25,600 + 14,400}}{6}$$

$$x = \frac{160 \pm \sqrt{40,000}}{6}$$

$$x = \frac{160 \pm 200}{6}$$

$$x = 60 \text{ mph} \quad \text{or} \quad \cancel{x = -\frac{20}{3}}$$

Problem Set 6.7, pp. 270-272

2. $x^2 + 9 = 0$

$x^2 = -9$

$x = \pm\sqrt{-9}$

$x = \pm 3i$

4. $x^2 + 16 = 0$

$x^2 = -16$

$x = \pm\sqrt{-16}$

$x = \pm 4i$

6. $x^2 + 36 = 0$

$x^2 = -36$

$x = \pm\sqrt{-36}$

$x = \pm 6i$

8. $x^2 + 3 = 0$

$x^2 = -3$

$x = \pm\sqrt{-3}$

$x = \pm\sqrt{3}i$

10. $y^2 + 7 = 0$

$$y^2 = -7$$

$$y = \pm\sqrt{-7}$$

$$y = \pm\sqrt{7}i$$

12. $p^2 + 20 = 0$

$$p^2 = -20$$

$$p = \pm\sqrt{-20}$$

$$p = \pm 2\sqrt{5}i$$

14. $r^2 + 63 = 0$

$$r^2 = -63$$

$$r = \pm\sqrt{-63}$$

$$r = \pm 3\sqrt{7}i$$

16. $9t^2 + 1 = 0$

$$t^2 = -\frac{1}{9}$$

$$t = \pm\sqrt{-\frac{1}{9}}$$

$$t = \pm\frac{1}{3}i$$

18. $49m^2 + 27 = 0$

$$49m^2 = -27$$

$$m^2 = -\frac{27}{49}$$

$$m = \pm\sqrt{\frac{-27}{49}}$$

$$m = \pm\frac{3\sqrt{3}i}{7}$$

20. $x^2 + 4x + 5 = 0$

$$x = \frac{-4 \pm \sqrt{4^2 - 4(1)(5)}}{2(1)}$$

$$x = \frac{-4 \pm \sqrt{16 - 20}}{2}$$

$$x = \frac{-4 \pm \sqrt{-4}}{2}$$

$$x = \frac{-4 \pm 2i}{2}$$

$$x = -2 \pm i$$

22. $r^2 - 6r + 10 = 0$

$$r = \frac{-(-6) \pm \sqrt{(-6)^2 - 4(1)(10)}}{2(1)}$$

$$r = \frac{6 \pm \sqrt{36 - 40}}{2}$$

$$r = \frac{6 \pm \sqrt{-4}}{2}$$

$$r = \frac{6 \pm 2i}{2}$$

$$r = 3 \pm i$$

24. $y^2 - 6y + 34 = 0$

$$y = \frac{-(-6) \pm \sqrt{(-6)^2 - 4(1)(34)}}{2(1)}$$

$$y = \frac{6 \pm \sqrt{36 - 136}}{2}$$

$$y = \frac{6 \pm \sqrt{-100}}{2}$$

$$y = \frac{6 \pm 10i}{2}$$

$$y = 3 \pm 5i$$

26. $2m^2 - 2m + 1 = 0$

$$m = \frac{-(-2) \pm \sqrt{(-2)^2 - 4(2)(1)}}{2(2)}$$

$$m = \frac{2 \pm \sqrt{4 - 8}}{4}$$

$$m = \frac{2 \pm \sqrt{-4}}{2}$$

$$m = \frac{2 \pm 2i}{2}$$

$$m = 1 \pm i$$

28. $z^2 + z + 1 = 0$

$$z = \frac{-1 \pm \sqrt{1^2 - 4(1)(1)}}{2(1)}$$

$$z = \frac{-1 \pm \sqrt{1 - 4}}{2}$$

$$z = \frac{-1 \pm \sqrt{-3}}{2}$$

$$z = \frac{-1 \pm \sqrt{3}i}{2}$$

30. $t^2 - 4t + 10 = 0$

$$t = \frac{-(-4) \pm \sqrt{(-4)^2 - 4(1)(10)}}{2(1)}$$

$$t = \frac{4 \pm \sqrt{16 - 40}}{2}$$

$$t = \frac{4 \pm \sqrt{-24}}{2}$$

$$t = \frac{4 \pm 2\sqrt{6}i}{2}$$

$$t = 2 \pm \sqrt{6}i$$

32. $x^4 + 21x^2 - 100 = 0$

$(x^2 + 25)(x^2 - 4) = 0$

$x^2 + 25 = 0$ or $x^2 - 4 = 0$

$$x^2 = -25 \qquad x^2 = 4$$

$$x = \pm\sqrt{-25} \qquad x = \pm\sqrt{4}$$

$$x = \pm 5i \qquad x = \pm 2$$

34. $y^4 - 2y^2 - 3 = 0$

$(y^2 - 3)(y^2 + 1) = 0$

$y^2 - 3 = 0$ or $y^2 + 1 = 0$

$$y^2 = 3 \qquad y^2 = -1$$

$$y = \pm\sqrt{3} \qquad y = \pm\sqrt{-1}$$

$$y = \pm i$$

36. $z^4 + 25z^2 + 144 = 0$

  $(z^2 + 9)(z^2 + 16) = 0$

  $z^2 + 9 = 0$ or $z^2 + 16 = 0$

  $z^2 = -9$ $z^2 = -16$

  $z = \pm\sqrt{-9}$ $z = \pm\sqrt{-16}$

  $z = \pm 3i$ $z = \pm 4i$

38. $x^2 + 5x + 4 = 0$

  $b^2 - 4ac = 5^2 - 4(1)(4) = 25 - 16 = 9$

  Since 9 is a perfect square, there are two rational solutions.

40. $4x^2 + 5x - 6 = 0$

  $b^2 - 4ac = 5^2 - 4(4)(-6) = 25 + 96 = 121$

  Since 121 is a perfect square, there are two rational solutions.

42. $x^2 + x - 9 = 0$

  $b^2 - 4ac = 1^2 - 4(1)(-9) = 1 + 36 = 37$

  Since 37 is positive but not a perfect square, there are two irrational solutions.

44. $m^2 = 2m - 1$

  $m^2 - 2m + 1 = 0$

  $b^2 - 4ac = (-2)^2 - 4(1)(1) = 4 - 4 = 0$

  Since the discriminant is 0, there is one rational solution.

46. $2m^2 + 3 = 4m$

  $2m^2 - 4m + 3 = 0$

  $b^2 - 4ac = (-4)^2 - 4(2)(3) = 16 - 24 = -8$

  Since -8 is negative, there are two complex solutions.

48. $$9m^2 = 12m - 4$$

$$9m^2 - 12m + 4 = 0$$

$$b^2 - 4ac = (-12)^2 - 4(9)(4) = 144 - 144 = 0$$

Since the discriminant is 0, there is one rational solution.

50. $x^2 + 8x + k = 0$

$$b^2 - 4ac = 0$$

$$8^2 - 4(1)k = 0$$

$$64 - 4k = 0$$

$$-4k = -64$$

$$k = 16$$

52. $2x^2 - kx + 2 = 0$

$$b^2 - 4ac = 0$$

$$(-k)^2 - 4(2)(2) = 0$$

$$k^2 - 16 = 0$$

$$k^2 = 16$$

$$k = \pm 4$$

54. $x + 2\left(\frac{1}{x}\right) = -2$

$$x + \frac{2}{x} = -2$$

$$x\left(x + \frac{2}{x}\right) = x(-2)$$

$$x^2 + 2 = -2x$$

$$x^2 + 2x + 2 = 0$$

$$x = \frac{-2 \pm \sqrt{2^2 - 4(1)(2)}}{2(1)}$$

$$x = \frac{-2 \pm \sqrt{4 - 8}}{2}$$

$$x = \frac{-2 \pm \sqrt{-4}}{2}$$

$$x = \frac{-2 \pm 2i}{2}$$

$$x = -1 \pm i$$

56. $$h = -16t^2 + 64t$$

$$128 = -16t^2 + 64t$$

$$16t^2 - 64t + 128 = 0$$

$$t^2 - 4t + 8 = 0$$

$$t = \frac{-(-4) \pm \sqrt{(-4)^2 - 4(1)(8)}}{2(1)}$$

$$t = \frac{4 \pm \sqrt{16 - 32}}{2}$$

$$t = \frac{4 \pm \sqrt{-16}}{2}$$

$$t = \frac{4 \pm 4i}{2}$$

$$t = 2 \pm 2i$$

Since the solutions are imaginary numbers, the arrow will never be at a height of 128 ft.

58. $25x^2 - 2x + 74 = 0$

$$x = \frac{-(-2) \pm \sqrt{(-2)^2 - 4(25)(74)}}{2(25)}$$

$$x = \frac{2 \pm \sqrt{4 - 7400}}{50}$$

$$x = \frac{2 \pm \sqrt{-7396}}{50}$$

$$x = \frac{2 \pm 86i}{50} = \frac{2(1 \pm 43i)}{50} = \frac{1 \pm 43i}{25}$$

Problem Set 6.8, pp. 277-278

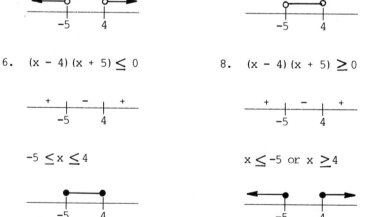

2. $(x - 4)(x + 5) > 0$

$x < -5$ or $x > 4$

4. $(x - 4)(x + 5) < 0$

$-5 < x < 4$

6. $(x - 4)(x + 5) \leq 0$

$-5 \leq x \leq 4$

8. $(x - 4)(x + 5) \geq 0$

$x \leq -5$ or $x \geq 4$

10. $y^2 - 5y + 4 \geq 0$

    $(y - 1)(y - 4) \geq 0$

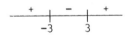

    $y \leq 1$ or $y \geq 4$

12. $z^2 + 6z + 5 > 0$

    $(z + 1)(z + 5) > 0$

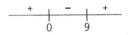

    $z < -5$ or $z > -1$

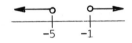

14. $x^2 - 9 < 0$

    $(x + 3)(x - 3) < 0$

    $-3 < x < 3$

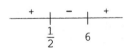

16. $x^2 - 9x < 0$

    $x(x - 9) < 0$

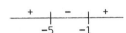

    $0 < x < 9$

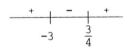

18. $2p^2 - 13p + 6 < 0$

    $(2p - 1)(p - 6) < 0$

    $\frac{1}{2} < p < 6$

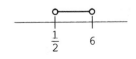

20. $4y^2 + 9y - 9 > 0$

    $(4y - 3)(y + 3) > 0$

    $y < -3$ or $y > \frac{3}{4}$

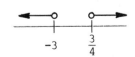

22. $m^2 + 5m \geq 0$

$m(m + 5) \geq 0$

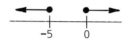

$m \leq -5$ or $m \geq 0$

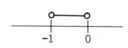

24. $t^2 - 16 > 0$

$(t + 4)(t - 4) > 0$

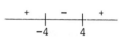

$t < -4$ or $t > 4$

26. $r^2 + r \leq 0$

$r(r + 1) \leq 0$

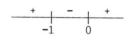

$-1 \leq r \leq 0$

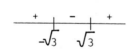

28. $4x^2 - 1 \geq 0$

$(2x + 1)(2x - 1) \geq 0$

$x \leq -\dfrac{1}{2}$ or $x \geq \dfrac{1}{2}$

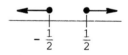

30. $y^2 - 3 \leq 0$

$(y + \sqrt{3})(y - \sqrt{3}) \leq 0$

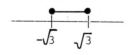

$-\sqrt{3} \leq y \leq \sqrt{3}$

32. $m^2 \geq 49$

$m^2 - 49 \geq 0$

$(m + 7)(m - 7) \geq 0$

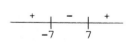

$m \leq -7$ or $m \geq 7$

34. $\quad p^2 \le 4p$

$p^2 - 4p \le 0$

$p(p - 4) \le 0$

$0 \le p \le 4$

36. $\quad y(y - 4) < y - 4$

$y^2 - 4y < y - 4$

$y^2 - 5y + 4 < 0$

$(y - 1)(y - 4) < 0$

(-4)

$1 < y < 4$

38. $\quad 6y(y - 3) + y > 7 + 2y$

$6y^2 - 18y + y > 7 + 2y$

$6y^2 - 19y - 7 > 0$

$(3y + 1)(2y - 7) > 0$

$y < -\dfrac{1}{3}$ or $y > \dfrac{7}{2}$

40. $\quad x^2 - 2x - 4 \ge 0$

$x = \dfrac{-(-2) \pm \sqrt{(-2)^2 - 4(1)(-4)}}{2(1)}$

$x = \dfrac{2 \pm \sqrt{4 + 16}}{2}$

$x = \dfrac{2 \pm \sqrt{20}}{2}$

$x = \dfrac{2 \pm 2\sqrt{5}}{2}$

$x = 1 \pm \sqrt{5}$

$1 - \sqrt{5} \approx 1.2 \quad 1 + \sqrt{5} \approx 3.2$

$x \le 1 - \sqrt{5}$ or $x \ge 1 + \sqrt{5}$

42. $\quad 2r^2 + 4r + 1 \le 0$

$r = \dfrac{-4 \pm \sqrt{4^2 - 4(2)(1)}}{2(2)}$

$r = \dfrac{-4 \pm \sqrt{16 - 8}}{4}$

$r = \dfrac{-4 \pm \sqrt{8}}{4} = \dfrac{-4 \pm 2\sqrt{2}}{4} = \dfrac{2(-2 \pm \sqrt{2})}{4} = \dfrac{-2 \pm \sqrt{2}}{2}$

$$\frac{-2-\sqrt{2}}{2} \approx -1.7 \qquad \frac{-2+\sqrt{2}}{2} \approx -0.3$$

$$\frac{-2-\sqrt{2}}{2} \leq r \leq \frac{-2+\sqrt{2}}{2}$$

44. $x^2 - 10x + 25 > 0$

    $(x - 5)^2 > 0$

    $x < 5$ or $x > 5$

46. $x^2 - 10x + 25 \geq 0$

    $(x - 5)^2 \geq 0$

    All real numbers

48. $x^2 - 10x + 25 < 0$

    $(x - 5)^2 < 0$

    No solution

50. $x^2 - 10x + 25 \leq 0$

    $(x - 5)^2 \leq 0$

    Only solution is 5

52. $x^2 + x + 3 > 0$

    $x = \dfrac{-1 \pm \sqrt{1^2 - 4(1)(3)}}{2(1)}$

    $x = \dfrac{-1 \pm \sqrt{1 - 12}}{2}$

    $x = \dfrac{-1 \pm \sqrt{-11}}{2}$

    All real numbers

54. $x^2 + x + 3 < 0$

    $x = \dfrac{-1 \pm \sqrt{1^2 - 4(1)(3)}}{2(1)}$

    $x = \dfrac{-1 \pm \sqrt{1 - 12}}{2}$

    $x = \dfrac{-1 \pm \sqrt{-11}}{2}$

    No solution

56. $y^2 + 9 \geq 0$

Solution is all real numbers since $y^2 \geq 0$.

58. $y^2 + 9 \leq 0$

No solution, since $y^2 \geq 0$.

60. $(r + 2)^2 > r^2 + 4r$

$r^2 + 4r + 4 > r^2 + 4r$

$4 > 0$

All real numbers

62. $(r + 2)^2 < r^2 + 4r$

$r^2 + 4r + 4 < r^2 + 4r$

$4 < 0$

No solution

64. $(y + 1)(y - 3)^2 > 0$

$-1 < y < 3$ or $y > 3$

66. $m^4 - 10m^2 + 9 \leq 0$

$(m^2 - 1)(m^2 - 9) \leq 0$

$(m - 1)(m + 1)(m - 3)(m + 3) \leq 0$

$-3 \leq m \leq -1$ or $1 \leq m \leq 3$

68. $h \geq 288$

$-16t^2 + 176t \geq 288$

$-16t^2 + 176t - 288 \geq 0$

$t^2 - 11t + 18 \leq 0$ Divide by -16

$(t - 2)(t - 9) \leq 0$

$2 \text{ sec} \leq t \leq 9 \text{ sec}$

70. $55y^2 - 288y + 377 > 0$

$$y = \frac{-(-288) \pm \sqrt{(-288)^2 - 4(55)(377)}}{2(55)}$$

$$y = \frac{288 \pm \sqrt{82,944 - 82,940}}{110}$$

$$y = \frac{288 \pm \sqrt{4}}{110} = \frac{288 \pm 2}{110}$$

$$y = \frac{288 + 2}{110} \quad \text{or} \quad y = \frac{288 - 2}{110}$$

$$y = \frac{29}{11} \qquad\qquad y = \frac{13}{5}$$

$$\begin{array}{c}
+ \quad\; | \quad - \quad | \quad + \\
\overline{\;\;\;\frac{13}{5}\quad\;\;\frac{29}{11}}
\end{array}$$

$y < \dfrac{13}{5}$ or $y > \dfrac{29}{11}$

CHAPTER 7

GRAPHING LINEAR EQUATIONS AND INEQUALITIES

Problem Set 7.1, pp. 287-288

2. through 16.

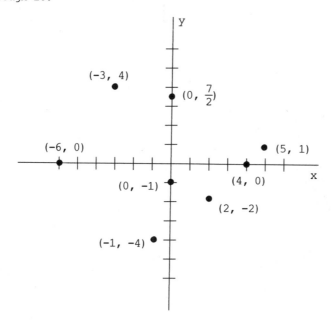

18. through 24.

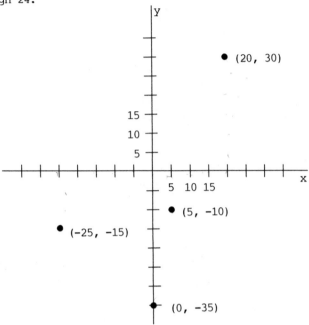

26. through 32.

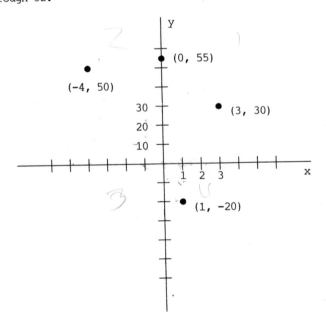

34. Since m is positive and n is negative, the point (n, m) = (-, +)
lies in quadrant II.

36. Since m is positive , -m is negative. Since n is negative, -n
is positive. Therefore (-n, -m) = (+,-) is in quadrant IV.

38. Since m is positive and n is negative, m - n is positive and
m/n is negative. Therefore (m - n, m/n) = (+, -) is in quadrant IV.

40. Since n is negative, the point (0, n) = (0, -) is on the
negative part of the y-axis.

42. There are five ways to roll an 8, namely (2, 6), (3, 5), (4, 4),
(5, 3), and (6, 2). Therefore P(8) = 5/36.

44. There are three ways to roll a 4, namely (1, 3), (2, 2), and
(3, 1). Therefore P(4) = 4/36 = 1/9.

46. There are six ways to roll a total less than 5, namely (1, 1),
(1, 2), (2, 1), (1, 3), (2, 2), and (3, 1). Therefore
P(<5) = 6/36 = 1/6.

48. There are thirty-six ways to roll a total greater than 0.
Therefore P(>0) = 36/36 = 1.

50. There are no ways to roll a 16. Therefore P(16) = 0/16 = 0.

52. A 2 or a 12, since there is only one way to roll each number

54. a)   b)

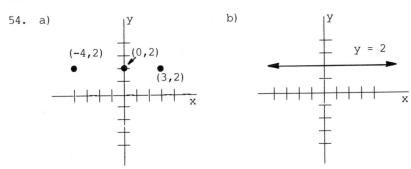

56.   a)

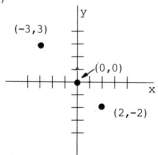

b)

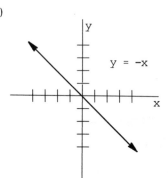

58.   There are eight ways to roll a 7 or 11, namely (1, 6), (2, 5), (3, 4), (4, 3), (5, 2), (6, 1), (5, 6), and (6, 5). Therefore P(7 or 11) = 8/36 = 2/9 = $0.22\overline{2}$ = $22.\overline{2}$%.

Problem Set 7.2, pp. 293-294

2.   y = 3x + 4 is linear, since it can be written as -3x + y = 4.

4. $y = x^2 - 1$ is not linear, because x has an exponent of 2.

6. $y = \sqrt[3]{x}$ is not linear, because $\sqrt[3]{x} = x^{1/3}$.

8. $y = \dfrac{8}{x}$ is not linear, because $\dfrac{8}{x} = 8x^{-1}$.

10.   y = x + 4

x	y
0	4
1	5
2	6
-4	0

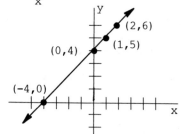

12.   y = 3x - 3

x	y
-1	-6
0	-3
1	0
7/3	4

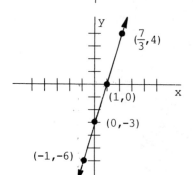

14.

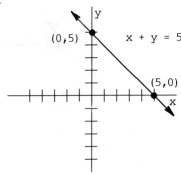

$x + y = 5$

16.

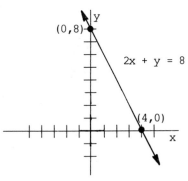

$2x + y = 8$

18.

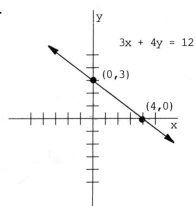

$3x + 4y = 12$

20.

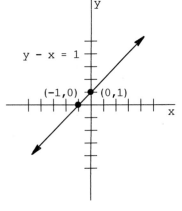

$y - x = 1$

22.

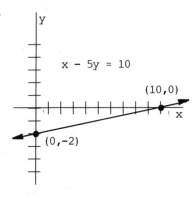

$x - 5y = 10$

24.

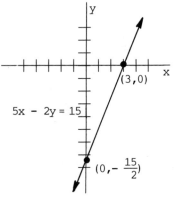

$5x - 2y = 15$

26.

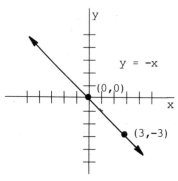

28.

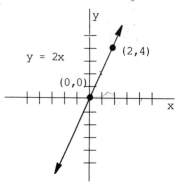

30.

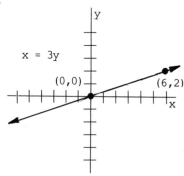

32.

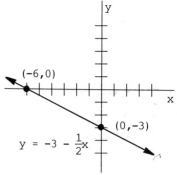

34. $y = 5$

x	y
-2	5
0	5
3	5
5	5

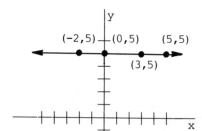

36. x = 1

x	y
1	-3
1	0
1	1
1	4

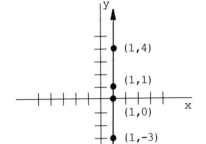

38.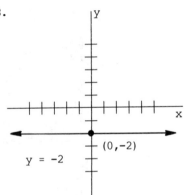

y = -2

40.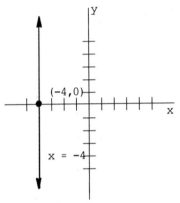

42. y - 6 = 0
     y = 6

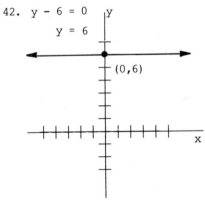

44. x - 4 = 0
     x = 4

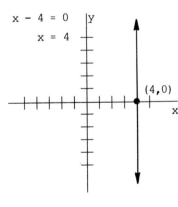

46.   2x + 7 = 0

$$x = -\frac{7}{2}$$

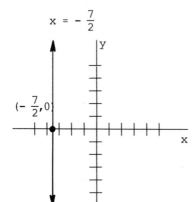

$\left(-\frac{7}{2}, 0\right)$

48.   2y - 3 = 0

$$y = \frac{3}{2}$$

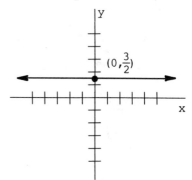

$\left(0, \frac{3}{2}\right)$

50.

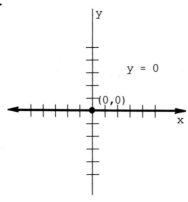

y = 0

(0,0)

52. $\frac{1}{4}x = 0$

x = 0

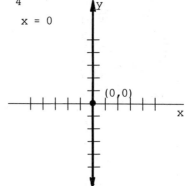

(0,0)

54.

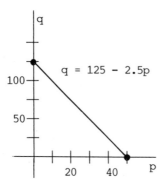

q = 125 - 2.5p

56.

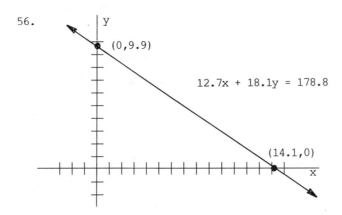

(0,9.9)

12.7x + 18.1y = 178.8

(14.1,0)

Problem Set 7.3, pp. 300-303

2. $d = \sqrt{(9 - 1)^2 + (8 - 2)^2} = \sqrt{8^2 + 6^2} = \sqrt{64 + 36} = \sqrt{100} = 10$

4. $d = \sqrt{(-3 - 2)^2 + (7 - (-5))^2} = \sqrt{(-5)^2 + 12^2} = \sqrt{25 + 144}$
$$= \sqrt{169}$$
$$= 13$$

6. $d = \sqrt{(2 - 1)^2 + (-2 - 1)^2} = \sqrt{1^2 + (-3)^2} = \sqrt{1 + 9} = \sqrt{10}$

8. $d = \sqrt{(-4 - 0)^2 + (1 - 0)^2} = \sqrt{(-4)^2 + 1^2} = \sqrt{16 + 1} = \sqrt{17}$

10. $d = \sqrt{(8 - 4)^2 + (11 - 7)^2} = \sqrt{4^2 + 4^2} = \sqrt{16 + 16} = \sqrt{32} = 4\sqrt{2}$

12. $d = \sqrt{(9 - 5)^2 + (-2 - 4)^2} = \sqrt{4^2 + (-6)^2} = \sqrt{16 + 36} = \sqrt{52}$
$$= 2\sqrt{13}$$

14. $d = \sqrt{(-1 - 2)^2 + (3 - (-6))^2} = \sqrt{(-3)^2 + 9^2} = \sqrt{9 + 81} = \sqrt{90}$
$$= 3\sqrt{10}$$

16. $d = \sqrt{(5.9 - 6.7)^2 + (-4.3 - (-2.8))^2} = \sqrt{(-0.8)^2 + (-1.5)^2}$
$$= \sqrt{0.64 + 2.25}$$
$$= \sqrt{2.89}$$
$$= 1.7$$

18. $d = \sqrt{(-\frac{1}{4} - (-\frac{1}{2}))^2 + (\frac{1}{3} - 0)^2} = \sqrt{(-\frac{1}{4} + \frac{2}{4})^2 + (\frac{1}{3})^2}$

$= \sqrt{\frac{1}{16} + \frac{1}{9}}$

$= \sqrt{\frac{9}{144} + \frac{16}{144}}$

$= \sqrt{\frac{25}{144}}$

$= \frac{5}{12}$

20. $d = \sqrt{(-ab - ab)^2 + (a^2 - b^2)^2} = \sqrt{(-2ab)^2 + a^4 - 2a^2b^2 + b^4}$

$= \sqrt{4a^2b^2 + a^4 - 2a^2b^2 + b^4}$

$= \sqrt{a^4 + 2a^2b^2 + b^4}$

$= \sqrt{(a^2 + b^2)^2}$

$= a^2 + b^2$

22. $d = \sqrt{(-3 - 5)^2 + (1 - 1)^2} = \sqrt{(-8)^2 + 0^2} = \sqrt{64 + 0} = \sqrt{64} = 8$

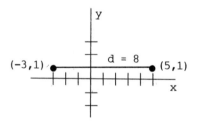

The two points are on the same horizontal line.

24. $d = \sqrt{(-5 - (-5))^2 + (-9 - (-2))^2} = \sqrt{0^2 + (-7)^2} = \sqrt{0 + 49}$

$= \sqrt{49}$

$= 7$

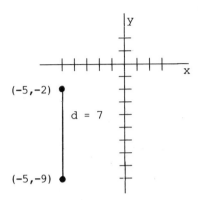

(-5,-2)

d = 7

(-5,-9)

The two points are on the same vertical line.

26. $m = \dfrac{9 - 6}{3 - 2} = \dfrac{3}{1} = 3$

28. $m = \dfrac{2 - (-8)}{-3 - (-1)} = \dfrac{10}{-2} = -5$

30. $m = \dfrac{-3 - 7}{-4 - 8} = \dfrac{-10}{-12} = \dfrac{5}{6}$

32. $m = \dfrac{-11 - (-11)}{5 - 4} = \dfrac{0}{1} = 0$

34. $m = \dfrac{-2 - 3}{1 - 1} = \dfrac{-5}{0}$ undefined

36. $m = \dfrac{6.5 - 7.8}{-4.2 - 1.9} = \dfrac{-1.3}{-6.1} = \dfrac{13}{61}$

38. $m = \dfrac{\frac{11}{12} - 0}{-\frac{1}{2} - (-\frac{1}{4})} = \dfrac{\frac{11}{12}}{-\frac{2}{4} + \frac{1}{4}} = \dfrac{\frac{11}{12}}{-\frac{1}{4}} = \dfrac{11}{12} \cdot (-\dfrac{4}{1}) = -\dfrac{11}{3}$

40. $m = \dfrac{(a + b) - (a - b)}{6 - 2} = \dfrac{a + b - a + b}{4} = \dfrac{2b}{4} = \dfrac{b}{2}$

42.

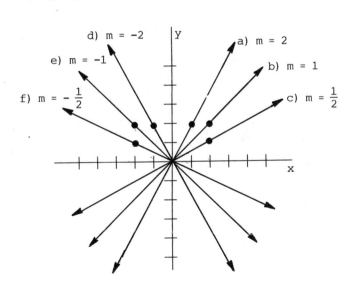

44.

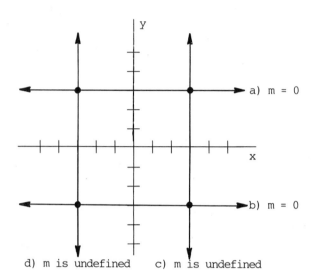

a) $m = 0$

b) $m = 0$

d) m is undefined   c) m is undefined

46. $m = \dfrac{4}{5}$

48. $m = \dfrac{-6}{2} = -3$

50. $m = 0$

52. m is undefined.

54. $AB = \sqrt{(1 - 0)^2 + (\sqrt{3} - 0)^2} = \sqrt{1^2 + (\sqrt{3})^2} = \sqrt{1 + 3} = \sqrt{4} = 2$

$AC = \sqrt{(-1 - 0)^2 + (\sqrt{3} - 0)^2} = \sqrt{(-1)^2 + (\sqrt{3})^2} = \sqrt{1 + 3} = \sqrt{4}$

$= 2$

$BC = \sqrt{(-1 - 1)^2 + (0 - 0)^2} = \sqrt{(-2)^2 + 0^2} = \sqrt{4 + 0} = \sqrt{4} = 2$

56. $5 = \sqrt{(5 - 1)^2 + (y - 4)^2}$

$5 = \sqrt{4^2 + y^2 - 8y + 16}$

$5 = \sqrt{y^2 - 8y + 32}$

Square both sides.

$25 = y^2 - 8y + 32$

$0 = y^2 - 8y + 7$

$0 = (y - 1)(y - 7)$

$y = 1$ or $y = 7$

58. $m_1 = \dfrac{-11 - (-5)}{12 - 2} = \dfrac{-6}{10} = -\dfrac{3}{5}$

a) $m_2 = m_1 = -\dfrac{3}{5}$

b) $m_2 = -\dfrac{1}{m_1} = \dfrac{5}{3}$

60. a) m_2 is undefined.

    b) $m_2 = 0$

62. $m_{AB} = \dfrac{5 - 2}{5 - 3} = \dfrac{3}{2}$ and $m_{CD} = \dfrac{8 - 5}{3 - 1} = \dfrac{3}{2}$.

    Therefore line AB is parallel to line CD.

    $m_{AD} = \dfrac{2 - 5}{3 - 1} = -\dfrac{3}{2}$ and $m_{BC} = \dfrac{5 - 8}{5 - 3} = -\dfrac{3}{2}$.

    Therefore line AD is parallel to line BC. Since m_{AB} and m_{AD} are not negative reciprocals, the parallelogram is not a square.

64. $\dfrac{t^2 - 3t}{9 - 2} = 4$

    $\dfrac{t^2 - 3t}{7} = 4$

    $t^2 - 3t = 28$

    $t^2 - 3t - 28 = 0$

    $(t - 7)(t + 4) = 0$

    $t = 7$ or $t = -4$

66. x = increase in elevation

6000 ft

$\dfrac{x}{6000} = 0.02$

$x = 120$ ft

68. $m = \dfrac{10.144 - 4.755}{20.288 - 28.213} = \dfrac{5.389}{-7.925} = -0.68$

Problem Set 7.4, pp. 307–309

2. $y - 4 = -3(x - 1)$

   $y - 4 = -3x + 3$

   $3x + y = 7$

4. $y - 8 = 2(x - (-5))$

   $y - 8 = 2x + 10$

   $-2x + y = 18$

   Multiply by -1.

   $2x - y = -18$

6. $y - (-2) = \frac{1}{4}(x - 6)$

$y + 2 = \frac{1}{4}(x - 6)$

$4y + 8 = x - 6$

$-x + 4y = -14$

$x - 4y = 14$

8. $y - (-7) = 1(x - (-3))$

$y + 7 = x + 3$

$-x + y = -4$

$x - y = 4$

10. $y - 11 = -\frac{5}{6}(x - 0)$

$6y - 66 = -5x$

$5x + 6y = 66$

12. $y - 3 = 0(x - 5)$

$y - 3 = 0$

$y = 3$

14. Since m is undefined, the line is vertical. Therefore the equation is $x = 4$.

16. $m = \frac{5 - 3}{5 - 4} = \frac{2}{1} = 2$

$y - 3 = 2(x - 4)$

$y - 3 = 2x - 8$

$-2x + y = -5$

$2x - y = 5$

18. $m = \frac{-7 - 0}{7 - 0} = \frac{-7}{7} = -1$

$y - 0 = -1(x - 0)$

$y = -x$

$x + y = 0$

20. $m = \frac{-9 - (-12)}{-2 - 4} = \frac{3}{-6} = -\frac{1}{2}$

$y - (-12) = -\frac{1}{2}(x - 4)$

$y + 12 = -\frac{1}{2}(x - 4)$

$2y + 24 = -(x - 4)$

$2y + 24 = -x + 4$

$x + 2y = -20$

22. $m = \frac{9 - 9}{5 - 2} = \frac{0}{3} = 0$

$y - 9 = 0(x - 2)$

$y - 9 = 0$

$y = 9$

24. $m = \frac{4 - (-1)}{-6 - (-6)} = \frac{5}{0}$, which is undefined.

Since m is undefined, the line is vertical. Therefore the equation is $x = -6$.

26. $m = \dfrac{-\sqrt{5} - \sqrt{5}}{2 - 6} = \dfrac{-2\sqrt{5}}{-4} = \dfrac{\sqrt{5}}{2}$

$y - \sqrt{5} = \dfrac{\sqrt{5}}{2}(x - 6)$

$2y - 2\sqrt{5} = \sqrt{5}(x - 6)$

$2y - 2\sqrt{5} = \sqrt{5}x - 6\sqrt{5}$

$-\sqrt{5}x + 2y = -4\sqrt{5}$

$\sqrt{5}x - 2y = 4\sqrt{5}$

28. $y = x + 2$

$m = 1, \; b = 2$

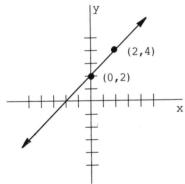

30. $y = 3x + 6$

$m = 3, \; b = 6$

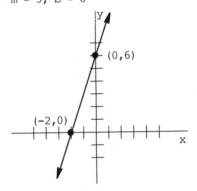

32. $5x + y = 3$

$y = -5x + 3$

$m = -5, \; b = 3$

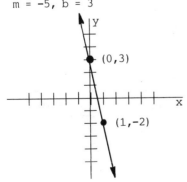

34. $3x - 4y = 8$

$-4y = -3x + 8$

$y = \dfrac{3}{4}x - 2$

$m = \dfrac{3}{4}, \; b = -2$

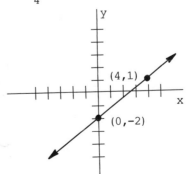

36. x + 2y = 5

$$2y = -x + 5$$

$$y = -\frac{1}{2}x + \frac{5}{2}$$

$$m = -\frac{1}{2}, \; b = \frac{5}{2}$$

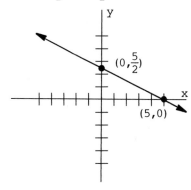

38. 5y - 4x = 0

$$5y = 4x$$

$$y = \frac{4}{5}x$$

$$m = \frac{4}{5}, \; b = 0$$

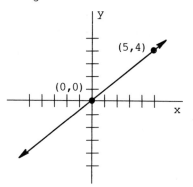

40. y - 4 = 0

$$y = 4$$

$$m = 0, \; b = 4$$

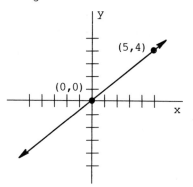

42. m = 1, b = -2

Equation is

$$y = 1 \cdot x + (-2)$$

$$y = x - 2.$$

44. $m = -\frac{2}{4} = -\frac{1}{2}, \; b = -2$

Equation is

$$y = -\frac{1}{2}x + (-2)$$

$$y = -\frac{1}{2}x - 2.$$

46. Since x = -4 is a vertical line, the slope is undefined and there is no y-intercept.

48.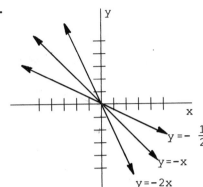

50.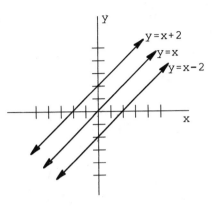

52. $y - x = 2$ $3x - 3y = 7$

$\qquad y = x + 2$ $\qquad -3y = -3x + 7$

$\qquad m = 1$ $\qquad\qquad y = x - \dfrac{7}{3}$

$\qquad\qquad\qquad m = 1$

The slopes are equal, so the lines are parallel.

54. $y = 5x - 1$ $2x + 10y = 0$

$\qquad m = 5$ $\qquad 10y = -2x$

$\qquad\qquad y = -\dfrac{1}{5}\,x$

$\qquad\qquad m = -\dfrac{1}{5}$

The slopes are negative reciprocals, so the lines are perpendicular.

56. $5x + 6y = 18$ $5x - 6y = -12$

$\qquad 6y = -5x + 18$ $\qquad -6y = -5x - 12$

$\qquad y = -\dfrac{5}{6}x + 3$ $\qquad y = \dfrac{5}{6}x + 2$

$\qquad m = -\dfrac{5}{6}$ $\qquad m = \dfrac{5}{6}$

The slopes are neither equal nor negative reciprocals, so the lines are neither parallel nor perpendicular.

58. $3x + 4y = 20$

$$4y = -3x + 20$$

$$y = -\frac{3}{4}x + 5$$

$$m = -\frac{3}{4}$$

a) $m = -\frac{3}{4}$, $b = \frac{3}{7}$

Equation is

$$y = -\frac{3}{4}x + \frac{3}{7}.$$

Multiply by 28.

$$28y = -21x + 12$$

$$21x + 28y = 12$$

b) $m = \frac{4}{3}$, $b = \frac{3}{7}$

Equation is

$$y = \frac{4}{3}x + \frac{3}{7}.$$

Multiply by 21.

$$21y = 28x + 9$$

$$-28x + 21y = 9$$

$$28x - 21y = -9$$

60.   Two points on the line are $(C,F) = (0,32)$ and $(C,F) = (100,212)$.
Therefore $b = 32$ and $m = \frac{212 - 32}{100 - 0} = \frac{180}{100} = \frac{9}{5}$.   The equation is
$F = \frac{9}{5}C + 32$.

62. $1.4579x + 3.05y = 26.047$

$$3.05y = -1.4579x + 26.047$$

$$y = -0.478x + 8.54$$

$$m = -0.478$$

$$2.99375x + 6.25y = 26.047$$

$$6.25y = -2.99375x + 26.047$$

$$y = -0.479x + 4.16752$$

$$m = -0.479$$

The slopes are neither equal nor negative reciprocals.   Therefore
the lines are neither parallel nor perpendicular.

Problem Set 7.5, pp. 314-315

2. x > 1

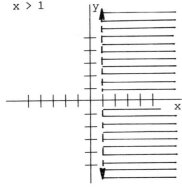

4. x ≥ 1

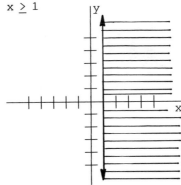

6. x < 1

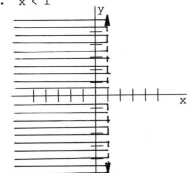

8. x ≤ 1

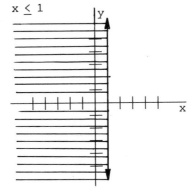

10. y ≤ 3

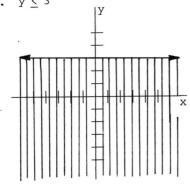

12. y + 4 > 0

     y > -4

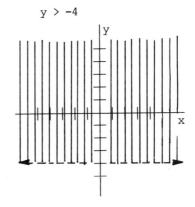

14. $2x + 5 \leq 0$

$$x \leq -\frac{5}{2}$$

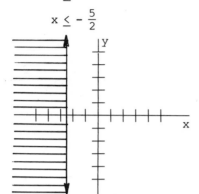

16.

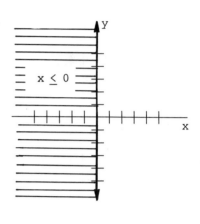

$x \leq 0$

18.

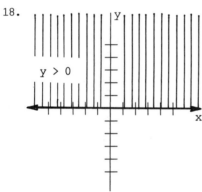

$y > 0$

y-axis not included

20. $y < x + 3$

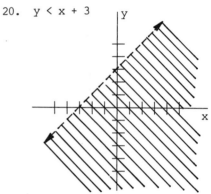

22. $y < 2x - 1$

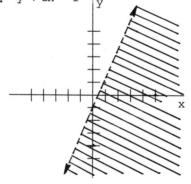

24. $x + y < 5$

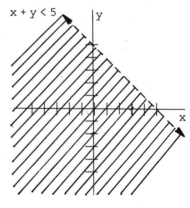

26. $4x - 5y \geq 20$

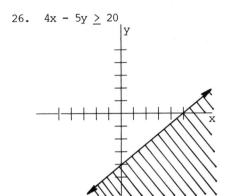

28. $2x - 3y \leq 6$

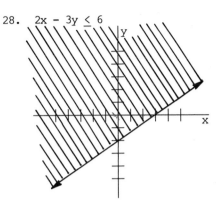

30. $y \geq 4x$

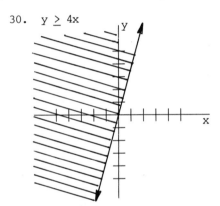

32. $x - y > 0$

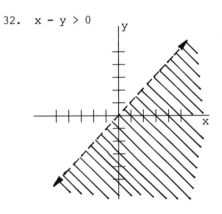

34. $x + 3y \geq 0$

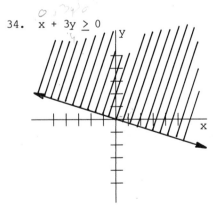

36. $-3 < x \leq 4$

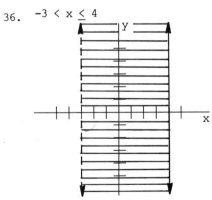

38. $-2 \le y < 1$

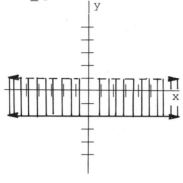

40. $|x| < 2$

      $-2 < x < 2$

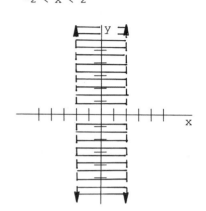

42. $|x| > 2$

      $x < -2$ or $x > 2$

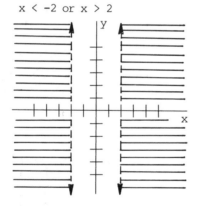

44. $|y| \le 4$

      $-4 \le y \le 4$

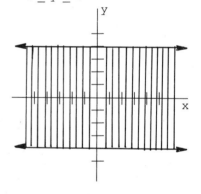

46. $|y| \ge 4$

      $y \le -4$ or $y \ge 4$

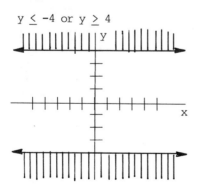

48. $|x - 3| \le 3$

      $-3 \le x - 3 \le 3$

         $0 \le x \le 6$

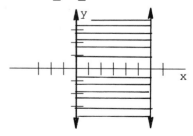

50. $|2y + 1| > 3$

$2y + 1 < -3$ or $2y + 1 > 3$

$2y < -4$ $2y > 2$

$y < -2$ $y > 1$

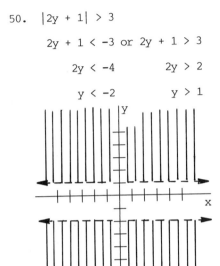

52. $4x + 8y \geq 24$

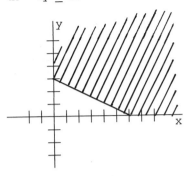

54. $|x| \leq 2$ and $|y| \leq 1$

$-2 \leq x \leq 2$ and $-1 \leq y \leq 1$

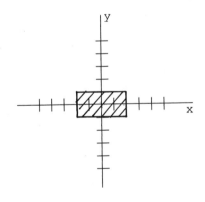

56. $|3.42x - 10.26| < 5.13$

$-5.13 < 3.42x - 10.26 < 5.13$

$5.13 < 3.42x < 15.39$

$1.5 < x < 4.5$

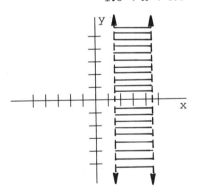

214

NOTES

CHAPTER 8

SYSTEMS OF LINEAR EQUATIONS

Problem Set 8.1, pp. 323-324

2. $y = 3x - 4$ $y = x + 2$

   $5 = 3(3) - 4$ $5 = 3 + 2$

   $5 = 5$ True $5 = 5$ True

Therefore $(3, 5)$ is a solution.

4. $5x - y = -9$ $3x - 6y = 27$

   $5(-1) - (-4) = -9$ $3(-1) - 6(-4) = 27$

       $-5 + 4 = -9$ $-3 + 24 = 27$

           $-1 = -9$ False $21 = 27$ False

Therefore $(-1, -4)$ is not a solution.

6. $2x - 7y = 26$ $x + 4y = 14$

   $2(6) - 7(-2) = 26$ $6 + 4(-2) = 14$

        $12 + 14 = 26$ $6 - 8 = 14$

           $26 = 26$ True $-2 = 14$ False

Therefore $(6, -2)$ is not a solution.

8.   Independent system

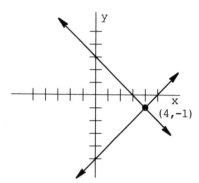

10.   Independent system

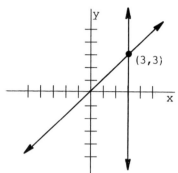

12.   Independent system

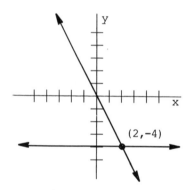

14.   Independent system

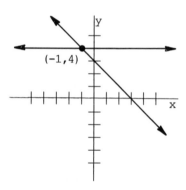

16.   Inconsistent System

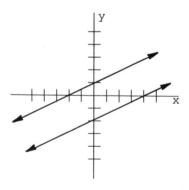

18.   Dependent system

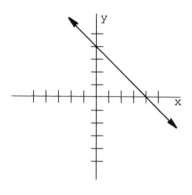

20. Independent system

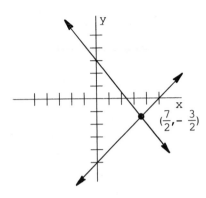

22. x + y = 5

x − y = 7

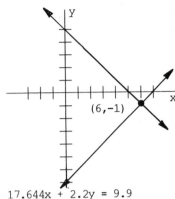

24. 12.03x + 1.5y = 6.6

1.5y = −12.03x + 6.6

y = −8.02x + 4.4

17.644x + 2.2y = 9.9

2.2y = −17.644x + 9.9

y = −8.02x + 4.5

The lines are parallel since they have the same slope, namely −8.02, but different y-intercepts (one is 4.4, the other is 4.5). Therefore the system is inconsistent.

Problem Set 8.2, pp. 328–329

2. 7x + y = 8

y = x

Replace y with x in the first equation.

7x + x = 8

8x = 8

x = 1

Replace x with 1 in the second equation.

y = 1

The solution is (1, 1).

4.   x = y + 2

x + 2y = 8

Replace x with y + 2 in the second equation.

(y + 2) + 2y = 8

3y = 6

y = 2

Replace y with 2 in the first equation.

x = 2 + 2

x = 4

The solution is (4, 2).

6. $3x + 4y = 19$

$y = 1$

Replace y with 1 in the first equation.

$3x + 4(1) = 19$

$3x = 15$

$x = 5$

The solution is $(5, 1)$.

8. $x = -2$

$2x + 5y = -4$

Replace x with -2 in the second equation.

$2(-2) + 5y = -4$

$-4 + 5y = -4$

$5y = 0$

$y = 0$

The solution is $(-2, 0)$.

10. $x + 3y = 16$

$4x - 5y = -38$

Solve the first equation for x.

$x = 16 - 3y$

Replace x with $16 - 3y$ in the second equation.

$4(16 - 3y) - 5y = -38$

$64 - 12y - 5y = -38$

$-17y = -102$

$y = 6$

Replace y with 6 in $x = 16 - 3y$.

$x = 16 - 3(6)$

$x = -2$

The solution is $(-2, 6)$.

12. $2x - y = 7$

$6x + y = 17$

Solve the second equation for y.

$y = 17 - 6x$

Replace y with $17 - 6x$ in the first equation.

$2x - (17 - 6x) = 7$

$2x - 17 + 6x = 7$

$8x = 24$

$x = 3$

Replace x with 3 in $y = 17 - 6x$.

$y = 17 - 6(3)$

$y = -1$

The solution is $(3, -1)$.

14. $3x + 10y = 5$

   $6x = 5y$

Solve the second equation for x.

$$x = \frac{5y}{6}$$

Replace x with 5y/6 in the first equation.

$$3\left(\frac{5y}{6}\right) + 10y = 5$$

$$\frac{5y}{2} + 10y = 5$$

Multiply by 2.

$5y + 20y = 10$

   $25y = 10$

$$y = \frac{2}{5}$$

Replace y with 2/5 in x = 5y/6.

$$x = \frac{5(2/5)}{6}$$

$$x = \frac{1}{3}$$

The solution is $\left(\frac{1}{3}, \frac{2}{5}\right)$.

16. $12x - 4y = 0$

   $y = 3x$

Replace y with 3x in the first equation.

$12x - 4(3x) = 0$

$12x - 12x = 0$

   $0 = 0$

This identity means the system is dependent. Any ordered pair that satisfies y = 3x is a solution.

18. $24x - 8y = 17$

   $-6x + 2y = 5$

Solve the second equation for y.

$2y = 5 + 6x$

$$y = \frac{5 + 6x}{2}$$

20. $5x + 4y = 1$

   $6x + 3y = 2$

Solve the second equation for y.

$3y = 2 - 6x$

$$y = \frac{2 - 6x}{3}$$

Replace y with (5 + 6x)/2 in the first equation.

$$24x - 8\left(\frac{5 + 6x}{2}\right) = 17$$

$$24x - 4(5 + 6x) = 17$$

$$24x - 20 - 24x = 17$$

$$-20 = 17$$

This contradiction means the system is inconsistent. There is no solution.

Replace y with (2 − 6x)/3 in the first equation.

$$5x + 4\left(\frac{2 - 6x}{3}\right) = 1$$

$$5x + \frac{8 - 24x}{3} = 1$$

Multiply by 3.

$$15x + 8 - 24x = 3$$

$$-9x = -5$$

$$x = \frac{5}{9}$$

Replace x with 5/9 in y = (2 − 6x)/3.

$$y = \frac{2 - 6(5/9)}{3}$$

$$y = \frac{2 - 10/3}{3}$$

$$y = \frac{-4/3}{3}$$

$$y = -\frac{4}{9}$$

The solution is $\left(\frac{5}{9}, -\frac{4}{9}\right)$.

22.   x + y = 9

4x − y = 6

Add the equations and get

$$5x = 15$$

$$x = 3.$$

Replace x with 3 in x + y = 9.

$$3 + y = 9$$

$$y = 6$$

The solution is (3, 6).

24.   2x − 8y = 2

9x + 8y = −13

Add the equations and get

$$11x = -11$$

$$x = -1.$$

Replace x with −1 in 9x + 8y = −13.

$$9(-1) + 8y = -13$$

$$8y = -4$$

$$y = -\frac{1}{2}$$

The solution is $\left(-1, -\frac{1}{2}\right)$.

26. $-x + 4y = -3$

   $2x - 3y = 6$

Multiply the first equation by 2 and add the result to the second equation.

   $-2x + 8y = -6$

   $\underline{2x - 3y = 6}$

   $5y = 0$

   $y = 0$

Replace y with 0 in the second equation.

   $2x - 3(0) = 6$

   $2x = 6$

   $x = 3$

The solution is (3, 0).

28. $3x + 4y = 19$

   $6x + 2y = 32$

Multiply the first equation by −2 and add the result to the second equation.

   $-6x - 8y = -38$

   $\underline{6x + 2y = 32}$

   $-6y = -6$

   $y = 1$

Replace y with 1 in the first equation.

   $3x + 4(1) = 19$

   $3x = 15$

   $x = 5$

The solution is (5, 1).

30. $3x - 5y = 6$

   $7x + 3y = -8$

Multiply the first equation by 3 and the second by 5.

   $9x - 15y = 18$

   $\underline{35x + 15y = -40}$

   $44x \qquad = -22$

   $x = -\dfrac{1}{2}$

Replace x with −1/2 in the second equation.

   $7(-\dfrac{1}{2}) + 3y = -8$

   $-\dfrac{7}{2} + 3y = -8$

Multiply by 2.

32. $x - 3y = 2$

   $-4x + 12y = -8$

Multiply the first equation by 4 and add the result to the second equation.

   $4x - 12y = 8$

   $\underline{-4x + 12y = -8}$

   $0 = 0$

This identity means the system is dependent. Any ordered pair that satisfies $x - 3y = 2$ is a solution.

$$-7 + 6y = -16$$

$$6y = -9$$

$$y = -\frac{3}{2}$$

The solution is $(-\frac{1}{2}, -\frac{3}{2})$.

34. $8x - 10y = 2$

$-12x + 15y = -1$

Multiply the first equation by 15 and the second by 10.

$120x - 150y = 30$

$\underline{-120x + 150y = -10}$

$0 = 20$

This contradiction means the system is inconsistent.   There is no solution.

36. $\frac{x}{3} - \frac{y}{4} = \frac{1}{3}$

$5x - 2y - 19 = 0$

Multiply the first equation by 12 and write the second equation in standard form.

$4x - 3y = 4$

$5x - 2y = 19$

Multiply $4x - 3y = 4$ by 2 and $5x - 2y = 19$ by -3.

$8x - 6y = 8$

$\underline{-15x + 6y = -57}$

$-7x \qquad = -49$

$x = 7$

Replace x with 7 in $4x - 3y = 4$.

$4(7) - 3y = 4$

$-3y = -24$

$y = 8$

The solution is $(7, 8)$.

38. $x + y = 21$

$x - y = 5$

Add the equations and get

$2x = 26$

$x = 13.$

40. $x + 3y = 22$

$x = y - 2$

Replace x with $y - 2$ in the first equation.

Replace x with 13 in the first equation.

13 + y = 21

y = 8

The numbers are 13 and 8.

$y - 2 + 3y = 22$

$4y = 24$

$y = 6$

Replace y with 6 in the second equation.

$x = 6 - 2$

$x = 4$

The numbers are 4 and 6.

42. $\dfrac{3}{x} + \dfrac{4}{y} = \dfrac{5}{2}$

$\dfrac{5}{x} - \dfrac{3}{y} = \dfrac{7}{4}$

Write as follows:

$3\left(\dfrac{1}{x}\right) + 4\left(\dfrac{1}{y}\right) = \dfrac{5}{2}$

$5\left(\dfrac{1}{x}\right) - 3\left(\dfrac{1}{y}\right) = \dfrac{7}{4}.$

Replace 1/x with u and 1/y with v.

$3u + 4v = \dfrac{5}{2}$

$5u - 3v = \dfrac{7}{4}$

Multiply the first equation by 2 and the second by 4.

$6u + 8v = 5$

$20u - 12v = 7$

Multiply the first equation by 12 and the second by 8.

$72u + 96v = 60$

$\underline{160u - 96v = 56}$

$232u \qquad = 116$

$u = \dfrac{1}{2}$

44. ax + by = c

x - y = 0

Solve the second equation for x.

x = y

Replace x with y in the first equation.

ay + by = c

(a + b)y = c

$y = \dfrac{c}{a + b}$

Replace y with c/(a + b) in x = y.

$x = \dfrac{c}{a + b}$

Replace u with 1/2 in
6u + 8v = 5.

$$6\left(\frac{1}{2}\right) + 8v = 5$$

$$3 + 8v = 5$$

$$8v = 2$$

$$v = \frac{1}{4}$$

Replace u with 1/x and
v with 1/y.

$$\frac{1}{x} = \frac{1}{2} \text{ and } \frac{1}{y} = \frac{1}{4}$$

Cross multiply in both equations.

$$2 = x \quad \text{and} \quad 4 = y$$

The solution is (2, 4).

46. 29.173x + 37.986y = 88.736

42.654x − 18.112y = 14.504

Multiply the first equation by 18.112 and the second by 37.986.

528.381x + 688.002y ≈ 1607.186

1620.255x − 688.002y ≈ 550.949
_____

2148.636x   ≈ 2158.135

$$x \approx 1.004$$

Replace x with 1.004 in the first equation.

29.173(1.004) + 37.986y ≈ 88.736

29.290 + 37.986y ≈ 88.736

37.986y ≈ 59.446

$$y \approx 1.56$$

The solution is (1.00, 1.56).

Problem Set 8.3, pp. 334-335

2. x = cost of ballpoint

y = cost of fountain pen

$1x + 3y = 29$

$5x + 2y = 28$

Multiply the first equation by -5.

$-5x - 15y = -145$

$\underline{5x + 2y = \qquad 28}$

$-13y = -117$

$y = \$9$

Substitute in x + 3y = 29.

$x + 3(9) = 29$

$x = \$2$

4. x = oz of potatoes

y = oz of meatloaf

$2x + 5y = 31$

$4x + 3y = 27$

Multiply the first equation by -2.

$-4x - 10y = -62$

$\underline{4x + 3y = \qquad 27}$

$-7y = -35$

$y = 5 \text{ oz}$

Substitute in 2x + 5y = 31.

$2x + 5(5) = 31$

$2x = 6$

$x = 3 \text{ oz}$

6. x = no. of baseballs

y = no. of footballs

$3x + 5y = 85$

$4x + 7y = 115$

Multiply the first equation by 4 and the second by -3.

$12x + 20y = 340$

$\underline{-12x - 21y = -345}$

$-y = -5$

$y = 5 \text{ baseballs}$

8. x = side of square

y = side of triangle

$4x = 3y$

$x + y = 35$

Solve the first equation for x.

$$x = \frac{3y}{4}$$

Substitute in x + y = 35.

$$\frac{3y}{4} + y = 35$$

Substitute in $3x + 5y = 85$.

$3x + 5(5) = 85$

$3x = 60$

$x = 20$ footballs

Multiply by 4.

$3y + 4y = 140$

$7y = 140$

$y = 20$ cm

Substitute in $x = 3y/4$.

$$x = \frac{3(20)}{4}$$

$x = 15$ cm

10. p = speed of plane

w = speed of wind

Since $r \cdot t = d$, write

$(p - w) \cdot 5 = 300$

$(p + w) \cdot 3 = 330$.

Divide the first equation by 5 and the second by 3.

$p - w = 60$

$\underline{p + w = 110}$

$2p \quad = 170$

$p = 85$ mph

Substitute in $p + w = 110$.

$85 + w = 110$

$w = 25$ mph

12. t = tens digit

u = units digit

$t + u = 12$

$10u + t = 10t + u - 54$

Simplify the second equation.

$t + u = 12$

$-9t + 9u = -54$

Multiply the first equation by 9.

$9t + 9u = 108$

$\underline{-9t + 9u = -54}$

$18u = 54$

$u = 3$

Substitute in $t + u = 12$.

$t + 3 = 12$

$t = 9$

The original number is 93.

14. x = Donald's age

   y = Daisy's age

   x − 7 = 6(y − 7)

   x + 1 = 2(y + 1)

   Simplify both equations.

   x − 6y = −35

   x − 2y = 1

   Multiply x − 6y = −35 by −1.

   −x + 6y = 35

   x − 2y = 1
   _____

   4y = 36

   y = 9 yr

   Substitute in x − 2y = 1.

   x − 2(9) = 1

   x = 19 yr

18. D = 216 − 6p

   S = 10p

   Set S = D.

   10p = 216 − 6p

   16p = 216

   p = $13.5

   Substitute in S = 10p.

   S = 10(13.5)

   S = 135 units

16. x = no. of Paul's potatoes

   y = no. of Julia's potatoes

   3(x − 2) = y + 2

   x + 2 = y − 2

   Simplify both equations.

   3x − y = 8

   x − y = −4

   Multiply x − y = −4 by −1.

   3x − y = 8

   −x + y = 4
   _____

   2x   = 12

   x = 6 potatoes

   Substitute in −x + y = 4.

   −6 + y = 4

   y = 10 potatoes

20. D = 1750 − 35p

   S = 100 + 15p

   Set S = D.

   100 + 15p = 1750 − 35p

   50p = 1650

   p = $33

   Substitute in S = 100 + 15p.

   S = 100 + 15(33)

   S = 595 units

22.   x = oz of 41% alloy

y = oz of 67% alloy

$$x + y = 101$$

$$0.41x + 0.67y = \frac{7}{12}(101)$$

Multiply the first equation by −0.41.

$$-0.41x - 0.41y = -41.41$$

$$\underline{0.41x + 0.67y = 58.92}$$

$$0.26y = 17.51$$

$$y = 67.3 \text{ oz}$$

Substitute in x + y = 101.

$$x + 67.3 = 101$$

$$x = 33.7 \text{ oz}$$

Problem Set 8.4, pp. 340–342

2. $x - y + z = -3$ $3x + 5y - 2z = 24$ $4x - 7y - 3z = -10$

$3 - 1 + (-5) = -3$ $3(3) + 5(1) - 2(-5) = 24$ $4(3) - 7(1) - 3(-5) = -10$

$-3 = -3$ $24 = 24$ $20 = -10$

True   True   False

(3, 1, −5) is not a solution.

4. $-4x + 9y + z = -5$ $16x - y + 6z = 0$ $8y - 5z = 10$

$-4\left(\frac{3}{4}\right) + 9(0) + (-2) = -5$ $16\left(\frac{3}{4}\right) - 0 + 6(-2) = 0$ $8(0) - 5(-2) = 10$

$-3 + 0 - 2 = -5$ $12 - 0 - 12 = 0$ $10 = 10$

$-5 = -5$ $0 = 0$ True

True   True

$\left(\frac{3}{4}, 0, -2\right)$ is a solution.

6. Add the first two equations.

$3x - y + z = 1$

$\underline{x + 2y - z = 3}$

$4x + y \quad\ = 4$

Add this equation to the third equation.

$4x + y = 4$

$\underline{x - y = 6}$

$5x \quad\ = 10$

$\quad x = 2$

Substitute in $x - y = 6$.

$2 - y = 6$

$-y = 4$

$y = -4$

Substitute in $x + 2y - z = 3$.

$2 + 2(-4) - z = 3$

$-6 - z = 3$

$-z = 9$

$z = -9$

The solution is $(2, -4, -9)$.

8. Add the first two equations.

$-x + 3y + z = 3$

$\underline{2x - 2y - z = 0}$

$x + y \quad\ = 3$

Add the second and third equations.

$2x - 2y - z = 0$

$\underline{x + y + z = 5}$

$3x - y \quad\ = 5$

Add the two equations in x and y.

$x + y = 3$

$\underline{3x - y = 5}$

$4x \quad\ = 8$

$x = 2$

Substitute in $x + y = 3$.

$2 + y = 3$

$y = 1$

Substitute in $x + y + z = 5$.

$2 + 1 + z = 5$

$z = 2$

The solution is $(2, 1, 2)$.

10. Add the first and third equations.

$2x - y + z = 5$

$\underline{-2x - y + 2z = -6}$

$-2y + 3z = -1$

12. Add the first two equations.

$x + y + z = 8$

$\underline{x - y + 3z = 2}$

$2x \quad\ + 4z = 10$

$x + 2z = 5$

Add the second and third equations.

$$2x + 2y - z = 9$$
$$\underline{-2x - y + 2z = -6}$$
$$y + z = 3$$

Multiply this equation by 2 and add the result to $-2y + 3z = -1$.

$$2y + 2z = 6$$
$$\underline{-2y + 3z = -1}$$
$$5z = 5$$
$$z = 1$$

Substitute in $y + z = 3$.

$$y + 1 = 3$$
$$y = 2$$

Substitute in $2x - y + z = 5$.

$$2x - 2 + 1 = 5$$
$$2x = 6$$
$$x = 3$$

The solution is $(3, 2, 1)$.

14.   Add the first two equations.

$$2x - y + 6z = -2$$
$$\underline{3x - 2y - 6z = -8}$$
$$5x - 3y = -10$$

Multiply the second equation by 4 and add the result to the third.

$$4x - 4y + 12z = 8$$
$$\underline{5x + 4y + 2z = 21}$$
$$9x + 14z = 29$$

Multiply $x + 2z = 5$ by -7 and add the result to $9x + 14z = 29$.

$$-7x - 14z = -35$$
$$\underline{9x + 14z = 29}$$
$$2x = -6$$
$$x = -3$$

Substitute in $x + 2z = 5$.

$$-3 + 2z = 5$$
$$2z = 8$$
$$z = 4$$

Substitute in $x + y + z = 8$.

$$-3 + y + 4 = 8$$
$$y = 7$$

The solution is $(-3, 7, 4)$.

16.   Multiply the first equation by -2 and add the result to the second.

$$-2x + 16y - 8z = -40$$
$$\underline{2x + 7y - 6z = 14}$$
$$23y - 14z = -26$$

Multiply the third equation by 2 and add it to the first.

$$2x + .6y - 6z = -6$$
$$2x - y + 6z = -2$$
$$4x + 5y \qquad = -8$$

Solving the system

$$5x - 3y = -10$$
$$4x + 5y = -8$$

gives $x = -2$ and $y = 0$.

Substituting gives $z = \frac{1}{3}$.

The solution is $(-2, 0, \frac{1}{3})$.

Multiply the first equation by -3 and add the result to the third.

$$-3x + 24y - 12z = -60$$
$$3x - 5y + 2z = 26$$
$$19y - 10z = -34$$

Solving the system

$$23y - 14z = -26$$
$$19y - 10z = -34$$

gives $y = -6$ and $z = -8$.

Substituting gives $x = 4$.

The solution is $(4, -6, -8)$.

18. Add the second and third equations.

$$6x + y - 3z = 5$$
$$5x + 4y + 3z = 30$$
$$11x + 5y \qquad = 35$$

Solving the system

$$2x - 5y = 30$$
$$11x + 5y = 35$$

gives $x = 5$ and $y = -4$.

Substituting gives $z = 7$.

The solution is $(5, -4, 7)$.

20. Multiply the first equation by 4 and the third by -3.

$$12x + 4y - 8z = 8$$
$$-12x + 3y - 3z = 30$$
$$7y - 11z = 38$$

Solving the system

$$7y - 4z = 17$$
$$7y - 11z = 38$$

gives $y = 5/7$ and $z = -3$.

Substituting gives $x = -11/7$.

The solution is $(-\frac{11}{7}, \frac{5}{7}, -3)$

22. Multiply the first equation by 4 and the second by 5.

$$12x - 20y \qquad = -20$$
$$20y + 10z = -50$$
$$12x \qquad + 10z = -70$$

$$6x + 5z = -35$$

24. Multiply the second equation by 20 and add the result to the first.

$$-4x - 5y + 10z = -20$$
$$4x + 5y - 10z = 20$$
$$0 = 0$$

Solving the system

$6x + 5z = -35$

$7x - 9z = -26$

gives $x = -5$ and $y = -2$.

Substituting gives $z = -1$.

The solution is $(-5, -2, -1)$.

This identity means there is no unique solution.

26. Multiply the first equation by 4 and add the result to the second.

$4x + 8y - 12z = 4$

$\underline{-4x - 8y + 12z = 5}$

$ 0 = 9$

This contradiction means there is no unique solution.

28. Multiply the second equation by 2 and add the result to the first equation.

$x + 3y - 2z = 0$

$\underline{6x - 2y + 2z = 0}$

$7x + y = 0$

Multiply the second equation by 3 and add the result to the third equation.

$9x - 3y + 3z = 0$

$\underline{2x + 4y - 3z = 0}$

$11x + y = 0$

Solving the system

$7x + y = 0$

$11x + y = 0$

gives $x = 0$ and $y = 0$.

Substituting gives $z = 0$.

The solution is $(0, 0, 0)$.

30. Add the first equation to the third.

$x + 6y - z = 0$

$\underline{4x - y + z = 0}$

$5x + 5y = 0$

$x + y = 0$

32. Write the system in standard form.

$x - 2y + z = 2$

$x - 11y + 2z = -2$

$2x - 13y + 3z = 4$

Add the second equation to the third.

$$-3x + 2y - z = 0$$

$$\underline{4x - y + z = 0}$$

$$x + y \qquad = 0$$

Solving the system

$$x + y = 0$$

$$x + y = 0$$

gives the identity $0 = 0$. Therefore there is no unique solution.

Multiply the first equation by -1 and add the result to the second.

$$-x + 2y - z = -2$$

$$\underline{x - 11y + 2z = -2}$$

$$-9y + z = -4$$

Multiply the first equation by -2 and add the result to the third.

$$-2x + 4y - 2z = -4$$

$$\underline{2x - 13y + 3z = 4}$$

$$-9y + z = 0$$

Solving the system

$$-9y + z = -4$$

$$-9y + z = 0$$

gives a contradiction.

Therefore there is no unique solution.

34. Add the first two equations.

$$x + 8y + 4z = 57$$

$$\underline{-x - 5y - 4z = -27}$$

$$3y = 30$$

$$y = 10$$

Multiply the second equation by 3 and add the result to the third.

$$-3x - 15y - 12z = -81$$

$$\underline{3x + 4y + 6z = 25}$$

$$-11y - 6z = -56$$

36. Replace y with 4x in the first equation.

$$4x + z = 5$$

Multiply this equation by -1 and add the result to the second equation.

$$-4x - z = -5$$

$$\underline{x + z = 2}$$

$$-3x = -3$$

$$x = 1$$

Substitute y = 10.

$$-11(10) - 6z = -56$$
$$-6z = 54$$
$$z = -9$$

Substituting gives x = 13.

The solution is (13, 10, -9).

Substituting gives

y = 4   and   z = 1.

Therefore the solution is (1, 4, 1).

38.   x = first number

y = second number

z = third number

x + y + z = 6

x + y - z = -10

2x + y + z = 0

Solving this system gives x = -6, y = 4, and z = 8.

40.   x = cost of a muffin

y = cost of a pie

z = cost of a cake

x + 2y + 3z = 26

x + 3y + 2z = 20

x + 4y + 5z = 42

Solving this system gives x = \$3, y = \$1, and z = \$7.

42.   x = width of rectangle

y = length of rectangle

z = side of triangle

2x + 2y = 2(3z)

x + y + z = 100

4x = y + z

Solving this system gives x = 20 cm, y = 55 cm, and z = 25 cm.

44.   x = score on first test

y = score on second test

z = score on third test

$$\frac{x + y + z}{3} = 73$$

y = x + 4

z = y + 7

Solving the system gives x = 68, y = 72, and z = 79.

46.   x = no. of Larry's bananas

y = no. of Curly's bananas

z = no. of Moe's bananas

48.   h = hundreds digit

t = tens digit

u = units digit

$x - 1 = y + 1$

$z + 1 = 2y$

$z - 5 = x + 3$

Solving this system gives
$x = 13$, $y = 11$, and $z = 21$.

$h + t + u = 19$

$t = u + 3$

$(100h+10t+u)-(100u+10t+h)=99$

Solving this system gives
$h = 6$, $t = 8$, and $u = 5$.
Therefore the number is 685.

50. Multiply the first equation by 2.53 and the third by 3.76.

$2.4035x \qquad\qquad - 9.5128z = \quad 52.371$

$\underline{\qquad\quad - 20.868y + 9.5128z = -110.168}$

$2.4035x - 20.868y \qquad\quad = \ -57.797$

Solving the system

$2.4035x - 20.868y = -57.797$

$\quad 7.85x - 4.78y = 1.36$

gives $x = 2$ and $y = 3$. Substituting gives $z = -5$. Therefore
the solution is $(2, 3, -5)$.

Problem Set 8.5, pp. 346-347

2. $\begin{vmatrix} 8 & 2 \\ 5 & 3 \end{vmatrix} = 8\cdot 3 - 5\cdot 2 = 24 - 10 = 14$

4. $\begin{vmatrix} 4 & 6 \\ 3 & 2 \end{vmatrix} = 4\cdot 2 - 3\cdot 6 = 8 - 18 = -10$

6. $\begin{vmatrix} 6 & 3 \\ 4 & 2 \end{vmatrix} = 6\cdot 2 - 4\cdot 3 = 12 - 12 = 0$

8. $\begin{vmatrix} -3 & \frac{1}{3} \\ 9 & -10 \end{vmatrix} = (-3)(-10) - 9\cdot\frac{1}{3} = 30 - 3 = 27$

10. $\begin{vmatrix} 8 & -1 \\ -4 & -2 \end{vmatrix} = 8(-2) - (-4)(-1) = -16 - 4 = -20$

12. $\begin{vmatrix} -4 & -4 \\ 3 & 5 \end{vmatrix} = (-4)5 - 3(-4) = -20 + 12 = -8$

14.
$$\begin{vmatrix} \frac{4}{5} & \frac{2}{5} \\ -\frac{4}{5} & \frac{3}{5} \end{vmatrix} = \frac{4}{5} \cdot \frac{3}{5} - \left(-\frac{4}{5}\right)\frac{2}{5} = \frac{12}{25} + \frac{8}{25} = \frac{20}{25} = \frac{4}{5}$$

16.
$$\begin{vmatrix} 0 & 1 \\ 1 & 0 \end{vmatrix} = 0 \cdot 0 - 1 \cdot 1 = 0 - 1 = -1$$

18.
$$\begin{vmatrix} 0 & 0 \\ 1 & 1 \end{vmatrix} = 0 \cdot 1 - 1 \cdot 0 = 0 - 0 = 0$$

20.
$$\begin{vmatrix} y & x \\ x & y \end{vmatrix} = y \cdot y - x \cdot x = y^2 - x^2$$

22.   The minor of 2 is $\begin{vmatrix} 4 & 6 \\ 7 & 9 \end{vmatrix}$.

24.   The minor of 6 is $\begin{vmatrix} 1 & 2 \\ 7 & 8 \end{vmatrix}$.

26.   The minor of 7 is $\begin{vmatrix} 2 & 3 \\ 5 & 6 \end{vmatrix}$.

28.   The minor of 9 is $\begin{vmatrix} 1 & 2 \\ 4 & 5 \end{vmatrix}$.

30.
$$\begin{vmatrix} 3 & -1 & -2 \\ 5 & 0 & -4 \\ 2 & -3 & 6 \end{vmatrix} = 3\begin{vmatrix} 0 & -4 \\ -3 & 6 \end{vmatrix} - 5\begin{vmatrix} -1 & -2 \\ -3 & 6 \end{vmatrix} + 2\begin{vmatrix} -1 & -2 \\ 0 & -4 \end{vmatrix}$$

$$= 3(0 - 12) - 5(-6 - 6) + 2(4 - 0)$$

$$= -36 + 60 + 8$$

$$= 32$$

32.
$$\begin{vmatrix} 1 & 5 & 1 \\ -2 & 2 & 4 \\ -4 & 3 & -1 \end{vmatrix} = 1\begin{vmatrix} 2 & 4 \\ 3 & -1 \end{vmatrix} - (-2)\begin{vmatrix} 5 & 1 \\ 3 & -1 \end{vmatrix} + (-4)\begin{vmatrix} 5 & 1 \\ 2 & 4 \end{vmatrix}$$

$$= 1(-2 - 12) + 2(-5 - 3) - 4(20 - 2)$$

$$= -14 - 16 - 72$$

$$= -102$$

34. $\begin{vmatrix} 3 & -1 & -2 \\ 5 & 0 & -4 \\ 2 & -3 & 6 \end{vmatrix} = -5\begin{vmatrix} -1 & -2 \\ -3 & 6 \end{vmatrix} + 0\begin{vmatrix} 3 & -2 \\ 2 & 6 \end{vmatrix} - (-4)\begin{vmatrix} 3 & -1 \\ 2 & -3 \end{vmatrix}$

$= -5(-6 - 6) + 0 + 4(-9 + 2)$

$= 60 + 0 - 28$

$= 32$

36. $\begin{vmatrix} 0 & 0 & 5 \\ 5 & 9 & 4 \\ 4 & 8 & -6 \end{vmatrix} = 0\begin{vmatrix} 9 & 4 \\ 8 & -6 \end{vmatrix} - 0\begin{vmatrix} 5 & 4 \\ 4 & -6 \end{vmatrix} + 5\begin{vmatrix} 5 & 9 \\ 4 & 8 \end{vmatrix} = 0 - 0 + 5(40 - 36) = 20$

38. $\begin{vmatrix} -3 & 2 & 1 \\ -2 & 8 & 0 \\ 4 & -1 & 1 \end{vmatrix} = 1\begin{vmatrix} -2 & 8 \\ 4 & -1 \end{vmatrix} - 0\begin{vmatrix} -3 & 2 \\ 4 & -1 \end{vmatrix} + 1\begin{vmatrix} -3 & 2 \\ -2 & 8 \end{vmatrix} = 1(2 - 32) - 0 + 1(-24 + 4)$

$= -30 - 20$

$= -50$

40. $\begin{vmatrix} 4 & -1 & 2 \\ -4 & 1 & -3 \\ 1 & 1 & 2 \end{vmatrix} = -(-1)\begin{vmatrix} -4 & -3 \\ 1 & 2 \end{vmatrix} + 1\begin{vmatrix} 4 & 2 \\ 1 & 2 \end{vmatrix} - 1\begin{vmatrix} 4 & 2 \\ -4 & -3 \end{vmatrix}$

$= 1(-8 + 3) + 1(8 - 2) - 1(-12 + 8)$

$= -5 + 6 + 4$

$= 5$

42. $\begin{vmatrix} -5 & 1 & 2 \\ -3 & 1 & -1 \\ 6 & -2 & -2 \end{vmatrix} = -1\begin{vmatrix} -3 & -1 \\ 6 & -2 \end{vmatrix} + 1\begin{vmatrix} -5 & 2 \\ 6 & -2 \end{vmatrix} - (-2)\begin{vmatrix} -5 & 2 \\ -3 & -1 \end{vmatrix}$

$= -1(-6 + 6) + 1(10 - 12) + 2(5 + 6)$

$= 0 - 2 + 22$

$= 20$

44. $\begin{vmatrix} 2 & 1 & -1 \\ 3 & -1 & -4 \\ 6 & 3 & -3 \end{vmatrix} = 2\begin{vmatrix} -1 & -4 \\ 3 & -3 \end{vmatrix} - 1\begin{vmatrix} 3 & -4 \\ 6 & -3 \end{vmatrix} + (-1)\begin{vmatrix} 3 & -1 \\ 6 & 3 \end{vmatrix}$

$= 2(3 + 12) - 1(-9 + 24) - 1(9 + 6)$

$= 30 - 15 - 15$

$= 0$

46. $\begin{vmatrix} a & b & c \\ 1 & 2 & 3 \\ 1 & 1 & 0 \end{vmatrix} = 1\begin{vmatrix} b & c \\ 2 & 3 \end{vmatrix} - 1\begin{vmatrix} a & c \\ 1 & 3 \end{vmatrix} + 0\begin{vmatrix} a & b \\ 1 & 2 \end{vmatrix}$

$= 1(3b - 2c) - 1(3a - c) + 0$

$= -3a + 3b - c$

48. $\begin{vmatrix} 1 & 0 & 0 \\ 0 & 1 & 0 \\ 0 & 0 & 1 \end{vmatrix} = 1\begin{vmatrix} 1 & 0 \\ 0 & 1 \end{vmatrix} - 0\begin{vmatrix} 0 & 0 \\ 0 & 1 \end{vmatrix} + 0\begin{vmatrix} 0 & 1 \\ 0 & 0 \end{vmatrix}$

$= 1(1 - 0) - 0 + 0$

$= 1$

50. $\begin{vmatrix} x & 1 \\ 3x & 4 \end{vmatrix} = 8$

$4x - 3x = 8$

$x = 8$

52. $\begin{vmatrix} x-1 & -7 \\ x+2 & -4 \end{vmatrix} = x + 4$

$-4x + 4 + 7x + 14 = x + 4$

$3x + 18 = x + 4$

$2x = -14$

$x = -7$

54. $\begin{vmatrix} x & 3 & -5 \\ 1 & 0 & 2 \\ 4 & x & -1 \end{vmatrix} = 24$

$-1\begin{vmatrix} 3 & -5 \\ x & -1 \end{vmatrix} + 0\begin{vmatrix} x & -5 \\ 4 & -1 \end{vmatrix} - 2\begin{vmatrix} x & 3 \\ 4 & x \end{vmatrix} = 24$

$-1(-3 + 5x) + 0 - 2(x^2 - 12) = 24$

$3 - 5x - 2x^2 + 24 = 24$

$0 = 2x^2 + 5x - 3$

$0 = (2x - 1)(x + 3)$

$x = \frac{1}{2} \text{ or } x = -3$

56. a) $\begin{vmatrix} 0 & a & x \\ 0 & b & y \\ 0 & c & z \end{vmatrix} = 0\begin{vmatrix} b & y \\ c & z \end{vmatrix} - 0\begin{vmatrix} a & x \\ c & z \end{vmatrix} + 0\begin{vmatrix} a & x \\ b & y \end{vmatrix} = 0$

   b) $\begin{vmatrix} a & 0 & x \\ b & 0 & y \\ c & 0 & z \end{vmatrix} = -0\begin{vmatrix} b & y \\ c & z \end{vmatrix} + 0\begin{vmatrix} a & x \\ c & z \end{vmatrix} - 0\begin{vmatrix} a & x \\ b & y \end{vmatrix} = 0$

   c) $\begin{vmatrix} a & x & 0 \\ b & y & 0 \\ c & z & 0 \end{vmatrix} = 0\begin{vmatrix} b & y \\ c & z \end{vmatrix} - 0\begin{vmatrix} a & x \\ c & z \end{vmatrix} + 0\begin{vmatrix} a & x \\ b & y \end{vmatrix} = 0$

   If any column of a 3 x 3 determinant consists of zeros, the value of the determinant is zero.

58. $\begin{vmatrix} 1.06 & -2.68 & 0 \\ 4.93 & 3.47 & -5.26 \\ 6.55 & -7.94 & -8.13 \end{vmatrix}$

   $= 1.06\begin{vmatrix} 3.47 & -5.26 \\ -7.94 & -8.13 \end{vmatrix} - (-2.68)\begin{vmatrix} 4.93 & -5.26 \\ 6.55 & -8.13 \end{vmatrix} + 0\begin{vmatrix} 4.93 & 3.47 \\ 6.55 & -7.94 \end{vmatrix}$

   $= 1.06(-69.9755) + 2.68(-5.6279) + 0$

   $= -89.256802$

Problem Set 8.6, pp. 351 - 352

2. $D = \begin{vmatrix} 4 & 3 \\ 2 & 5 \end{vmatrix} = 14$

   $D_x = \begin{vmatrix} 11 & 3 \\ 9 & 5 \end{vmatrix} = 28$

   $D_y = \begin{vmatrix} 4 & 11 \\ 2 & 9 \end{vmatrix} = 14$

   $x = \dfrac{D_x}{D} = \dfrac{28}{14} = 2$

   $y = \dfrac{D_y}{D} = \dfrac{14}{14} = 1$

   The solution is (2, 1).

4. $D = \begin{vmatrix} 1 & 1 \\ 5 & 6 \end{vmatrix} = 1$

   $D_x = \begin{vmatrix} 1 & 1 \\ -2 & 6 \end{vmatrix} = 8$

   $D_y = \begin{vmatrix} 1 & 1 \\ 5 & -2 \end{vmatrix} = -7$

   $x = \dfrac{D_x}{D} = \dfrac{8}{1} = 8$

   $y = \dfrac{D_y}{D} = \dfrac{-7}{1} = -7$

   The solution is (8, -7).

6. $D = \begin{vmatrix} 8 & -3 \\ 4 & 6 \end{vmatrix} = 60$

$D_x = \begin{vmatrix} 7 & -3 \\ -9 & 6 \end{vmatrix} = 15$

$D_y = \begin{vmatrix} 8 & 7 \\ 4 & -9 \end{vmatrix} = -100$

$x = \dfrac{D_x}{D} = \dfrac{15}{60} = \dfrac{1}{4}$

$y = \dfrac{D_y}{D} = \dfrac{-100}{60} = -\dfrac{5}{3}$

The solution is $(\dfrac{1}{4}, -\dfrac{5}{3})$.

8. $D = \begin{vmatrix} 7 & -1 \\ -3 & 4 \end{vmatrix} = 25$

$D_x = \begin{vmatrix} 28 & -1 \\ -12 & 4 \end{vmatrix} = 100$

$D_y = \begin{vmatrix} 7 & 28 \\ -3 & -12 \end{vmatrix} = 0$

$x = \dfrac{D_x}{D} = \dfrac{100}{25} = 4$

$y = \dfrac{D_y}{D} = \dfrac{0}{25} = 0$

The solution is $(4, 0)$.

10. Multiply the first equation by 4 and the second by 2.

$4x = y - 8$

$2y = 8x + 5$

Write in standard form.

$4x - y = -8$

$-8x + 2y = 5$

$D = \begin{vmatrix} 4 & -1 \\ -8 & 2 \end{vmatrix} = 0$

Since $D = 0$, there is no unique solution.

12. $D = \begin{vmatrix} 0 & 1 \\ 8 & 5 \end{vmatrix} = -8$

$D_x = \begin{vmatrix} -2 & 1 \\ 14 & 5 \end{vmatrix} = -24$

$D_y = \begin{vmatrix} 0 & -2 \\ 8 & 14 \end{vmatrix} = 16$

$x = \dfrac{D_x}{D} = \dfrac{-24}{-8} = 3$

$y = \dfrac{D_y}{D} = \dfrac{16}{-8} = -2$

The solution is $(3, -2)$.

14. $D = \begin{vmatrix} 1 & 1 & 1 \\ 1 & 1 & -1 \\ 2 & -4 & -3 \end{vmatrix} = -12$

$D_x = \begin{vmatrix} 6 & 1 & 1 \\ 0 & 1 & -1 \\ 3 & -4 & -3 \end{vmatrix} = -48$

$D_y = \begin{vmatrix} 1 & 6 & 1 \\ 1 & 0 & -1 \\ 2 & 3 & -3 \end{vmatrix} = 12$

16. $D = \begin{vmatrix} 1 & -2 & 1 \\ 4 & -1 & 2 \\ 3 & 1 & 1 \end{vmatrix} = 0$

Since $D = 0$, there is no unique solution.

$$D_z = \begin{vmatrix} 1 & 1 & 6 \\ 1 & 1 & 0 \\ 2 & -4 & 3 \end{vmatrix} = -36$$

$$x = \frac{D_x}{D} = \frac{-48}{-12} = 4$$

$$y = \frac{D_y}{D} = \frac{12}{-12} = -1$$

$$z = \frac{D_z}{D} = \frac{-36}{-12} = 3$$

The solution is $(4, -1, 3)$.

18.
$$D = \begin{vmatrix} 1 & 1 & 0 \\ 0 & 2 & 5 \\ 2 & 0 & -10 \end{vmatrix} = -10$$

$$D_x = \begin{vmatrix} -1 & 1 & 0 \\ -8 & 2 & 5 \\ 4 & 0 & -10 \end{vmatrix} = -40$$

$$D_y = \begin{vmatrix} 1 & -1 & 0 \\ 0 & -8 & 5 \\ 2 & 4 & -10 \end{vmatrix} = 50$$

$$D_z = \begin{vmatrix} 1 & 1 & -1 \\ 0 & 2 & -8 \\ 2 & 0 & 4 \end{vmatrix} = -4$$

$$x = \frac{D_x}{D} = \frac{-40}{-10} = 4$$

$$y = \frac{D_y}{D} = \frac{50}{-10} = -5$$

$$z = \frac{D_z}{D} = \frac{-4}{-10} = \frac{2}{5}$$

The solution is $(4, -5, \frac{2}{5})$

20. Write in standard form.

$$4x \qquad - 3z = -3$$
$$x - 2y + 2z = 2$$
$$5y + z = 1$$

$$D = \begin{vmatrix} 4 & 0 & -3 \\ 1 & -2 & 2 \\ 0 & 5 & 1 \end{vmatrix} = -63$$

$$D_x = \begin{vmatrix} -3 & 0 & -3 \\ 2 & -2 & 2 \\ 1 & 5 & 1 \end{vmatrix} = 0$$

$$D_y = \begin{vmatrix} 4 & -3 & -3 \\ 1 & 2 & 2 \\ 0 & 1 & 1 \end{vmatrix} = 0$$

$$D_z = \begin{vmatrix} 4 & 0 & -3 \\ 1 & -2 & 2 \\ 0 & 5 & 1 \end{vmatrix} = -63$$

$$x = \frac{D_x}{D} = \frac{0}{-63} = 0$$

$$y = \frac{D_y}{D} = \frac{0}{-63} = 0$$

$$z = \frac{D_z}{D} = \frac{-63}{-63} = 1$$

The solution is $(0, 0, 1)$.

22.
$$D = \begin{vmatrix} 1 & -3 & 1 \\ 2 & -6 & 2 \\ -1 & 3 & -2 \end{vmatrix} = 0$$

Since D = 0, there is no unique solution.

24.
$$D = \begin{vmatrix} -1 & 2 & -1 \\ 3 & -1 & 4 \\ 5 & -3 & -1 \end{vmatrix} = 37$$

$$D_x = \begin{vmatrix} 0 & 2 & -1 \\ 0 & -1 & 4 \\ 0 & -3 & -1 \end{vmatrix} = 0$$

$$D_y = \begin{vmatrix} -1 & 0 & -1 \\ 3 & 0 & 4 \\ 5 & 0 & -1 \end{vmatrix} = 0$$

$$D_z = \begin{vmatrix} -1 & 2 & 0 \\ 3 & -1 & 0 \\ 5 & -3 & 0 \end{vmatrix} = 0$$

$$x = \frac{D_x}{D} = \frac{0}{37} = 0$$

$$y = \frac{D_y}{D} = \frac{0}{37} = 0$$

$$z = \frac{D_z}{D} = \frac{0}{37} = 0$$

The solution is (0, 0, 0).

26.
$$D = \begin{vmatrix} 161 & -30 & 17 \\ 43 & 79 & -22 \\ 101 & -88 & -99 \end{vmatrix} = -1,831,898$$

$$D_x = \begin{vmatrix} 527 & -30 & 17 \\ -340 & 79 & -22 \\ 59 & -88 & -99 \end{vmatrix} = -3,663,79$$

$$D_y = \begin{vmatrix} 161 & 527 & 17 \\ 43 & -340 & -22 \\ 101 & 59 & -99 \end{vmatrix} = 7,327,592$$

$$D_z = \begin{vmatrix} 161 & -30 & 527 \\ 43 & 79 & -340 \\ 101 & -88 & 59 \end{vmatrix} = -9,159,49$$

$$x = \frac{D_x}{D} = \frac{-3,663,796}{-1,831,898} = 2$$

$$y = \frac{D_y}{D} = \frac{7,327,592}{-1,831,898} = -4$$

$$z = \frac{D_z}{D} = \frac{-9,159,490}{-1,831,898} = 5$$

The solution is (2, -4, 5).

Problem Set 9.1, pp. 363-364

2. $y = x^2 - 9$

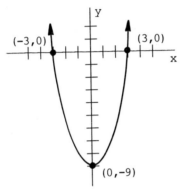

4. $y = 2x^2 - 2$

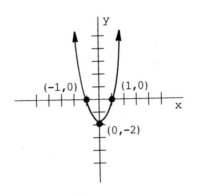

6. $y = -x^2 + 4$

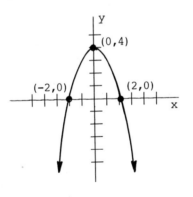

8. $y = 3 - x^2$

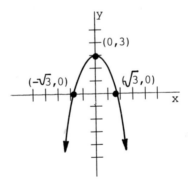

10. $y = x^2 + 2$

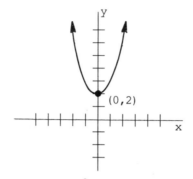

12. $y = x^2 + 4x$

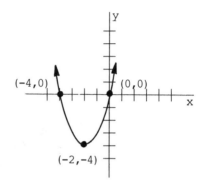

14. $y = 6x - 3x^2$

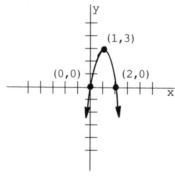

16. $y = x^2 - 4x + 3$

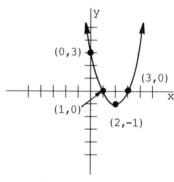

18. $y = x^2 - 2x - 8$

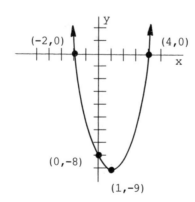

20. $y = -x^2 - 8x - 7$

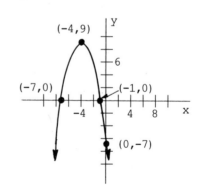

22. $y = -x^2 + 6x - 9$

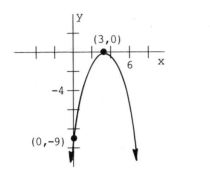

24. $y = 3x^2 + 2x - 1$

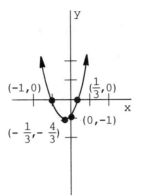

26. $y = x^2 - 3x + 1$

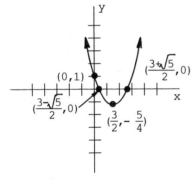

28. $y = \frac{1}{4}x^2 - 2x + 4$

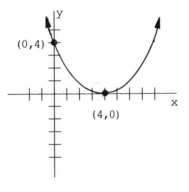

30. $y = \frac{1}{2}x^2 - x + 1$

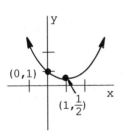

32.

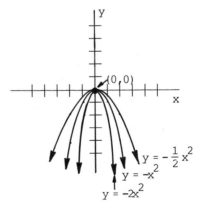

34.

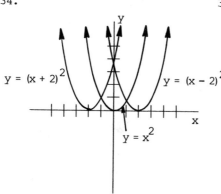

$y = (x + 2)^2$ $y = (x - 2)^2$

$y = x^2$

36.

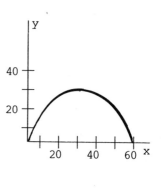

38.   A = cross-sectional area

A = x(16 - 2x)

$A = -2x^2 + 16x$

Maximum area occurs at

$$x = \frac{-b}{2a} = \frac{-16}{2(-2)} = 4 \text{ in.}$$

Maximum area is

$A = -2(4)^2 + 16(4) = 32 \text{ in.}^2$

40. $x = y^2 - 4$

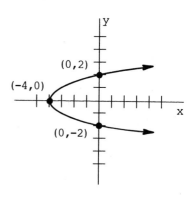

(0,2)

(-4,0)

(0,-2)

42. $x = -y^2 + 6y$

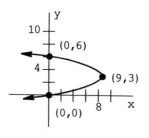

(0,6)

(9,3)

(0,0)

44. $P = -0.68x^2 + 176.8x - 6807$

Maximum profit occurs at

$$x = \frac{-b}{2a} = \frac{-176.8}{2(-0.68)} = 130 \text{ chairs.}$$

Maximum profit is

$P = -0.68(130)^2 + 176.8(130) - 68$

$P = 4685.$

Problem Set 9.2, pp. 367-368

2. $(x - 1)^2 + (y - 4)^2 = 25^2$

$(x - 1)^2 + (y - 4)^2 = 625$

4. $(x - 9)^2 + (y - (-5))^2 = 6^2$

$(x - 9)^2 + (y + 5)^2 = 36$

6. $(x - (-3))^2 + (y - (-\frac{1}{3}))^2 = 1^2$

$(x + 3)^2 + (y + \frac{1}{3})^2 = 1$

8. $(x - 2)^2 + (y - 0)^2 = (\frac{1}{3})^2$

$(x - 2)^2 + y^2 = \frac{1}{9}$

10. $(x - 0)^2 + (y - 0)^2 = (\sqrt{3})^2$

$x^2 + y^2 = 3$

12. $x^2 + y^2 = 16$

$(x - 0)^2 + (y - 0)^2 = 4^2$

$C(0, 0), r = 4$

14. $5x^2 + 5y^2 = 5$

Divide by 5.

$x^2 + y^2 = 1$

$(x - 0)^2 + (y - 0)^2 = 1^2$

$C(0, 0), r = 1$

16. $9x^2 + 9y^2 = 16$

$x^2 + y^2 = \frac{16}{9}$

$(x - 0)^2 + (y - 0)^2 = (\frac{4}{3})^2$

$C(0, 0), r = \frac{4}{3}$

18. $x^2 + (y - 2)^2 = 4$

$(x - 0)^2 + (y - 2)^2 = 2^2$

$C(0, 2), r = 2$

20. $(x - 5)^2 + y^2 = 100$

$(x - 5)^2 + (y - 0)^2 = 10^2$

$C(5, 0), r = 10$

22. $(x - 3)^2 + (y + 4)^2 = 64$

$(x - 3)^2 + (y - (-4))^2 = 8^2$

$C(3, -4), r = 8$

24. $(x + \frac{1}{5})^2 + (y + \frac{3}{5})^2 = \frac{1}{4}$

$(x - (-\frac{1}{5}))^2 + (y - (-\frac{3}{5}))^2 = (\frac{1}{2})^2$

$C(-\frac{1}{5}, -\frac{3}{5}), r = \frac{1}{2}$

26. $(x - 27)^2 + (y - 27)^2 = 27$

$(x - 27)^2 + (y - 27)^2 = (\sqrt{27})^2$

$C(27, 27), r = \sqrt{27} = 3\sqrt{3}$

28. $x^2 + y^2 - 6x + 4y - 3 = 0$

$(x^2 - 6x) + (y^2 + 4y) = 3$

$(x^2 - 6x + 9) + (y^2 + 4y + 4) = 3 + 9 + 4$

$(x - 3)^2 + (y + 2)^2 = 16$

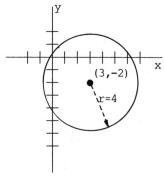

30. $x^2 + y^2 - 10x - 8y + 40 = 0$

$(x^2 - 10x) + (y^2 - 8y) = -40$

$(x^2 - 10x + 25) + (y^2 - 8y + 16) = -40 + 25 + 16$

$(x - 5)^2 + (y - 4)^2 = 1$

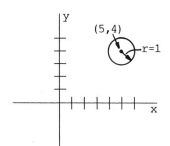

32.
$$x^2 + y^2 + 2x - 2y - 2 = 0$$
$$(x^2 + 2x) + (y^2 - 2y) = 2$$
$$(x^2 + 2x + 1) + (y^2 - 2y + 1) = 2 + 1 + 1$$
$$(x + 1)^2 + (y - 1)^2 = 4$$

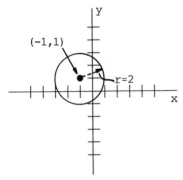

34.
$$x^2 + y^2 + 14x + 24 = 0$$
$$(x^2 + 14x) + y^2 = -24$$
$$(x^2 + 14x + 49) + y^2 = -24 + 49$$
$$(x + 7)^2 + (y - 0)^2 = 25$$

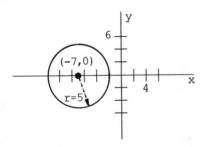

36.
$$x^2 + y^2 - 6x = 0$$
$$(x^2 - 6x) + y^2 = 0$$
$$(x^2 - 6x + 9) + y^2 = 0 + 9$$
$$(x - 3)^2 + (y - 0)^2 = 9$$

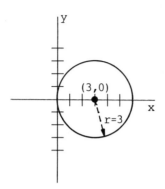

38. $x^2 + y^2 = 9$ 40.

$(x - 0)^2 + (y - 0)^2 = 9$

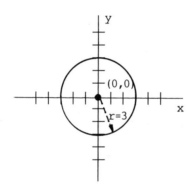

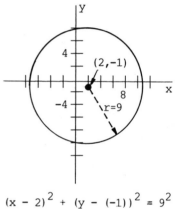

$(x - 2)^2 + (y - (-1))^2 = 9^2$

$(x - 2)^2 + (y + 1)^2 = 81$

42.   By the distance formula,

$$r = \sqrt{(5 - (-2))^2 + (-1 - (-4))^2}$$

$$r = \sqrt{49 + 9} = \sqrt{58} .$$

An equation for the circle is

$$(x - (-2))^2 + (y - (-4))^2 = (\sqrt{58})^2$$

$$(x + 2)^2 + (y + 4)^2 = 58.$$

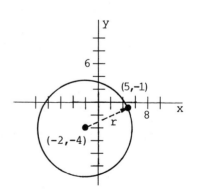

44. $9x^2 + 9y^2 - 6x + 12y + 1 = 0$

Divide by 9.

$$x^2 + y^2 - \frac{2}{3}x + \frac{4}{3}y + \frac{1}{9} = 0$$

$$(x^2 - \frac{2}{3}x) + (y^2 + \frac{4}{3}y) = -\frac{1}{9}$$

$$(x^2 - \frac{2}{3}x + \frac{1}{9}) + (y^2 + \frac{4}{3}y + \frac{4}{9}) = -\frac{1}{9} + \frac{1}{9} + \frac{4}{9}$$

$$(x - \frac{1}{3})^2 + (y + \frac{2}{3})^2 = \frac{4}{9}$$

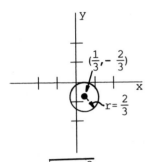

46. $y = -\sqrt{25 - x^2}$

x	y
5	0
4	-3
3	-4
0	-5
-3	-4
-4	-3
-5	0

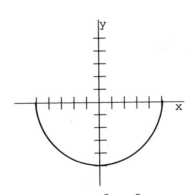

The graph is the bottom half of the circle $x^2 + y^2 = 25$.

48.
$$x^2 + y^2 + 6.4x - 3.9y - 4.02 = 0$$
$$(x^2 + 6.4x) + (y^2 - 3.9y) = 4.02$$
$$(x^2 + 6.4x + 10.24) + (y^2 - 3.9y + 3.8025) = 4.02 + 10.24 + 3.8025$$
$$(x + 3.2)^2 + (y - 1.95)^2 = 18.0625$$
$$C(-3.2, 1.95), \ r = \sqrt{18.0625} = 4.25$$

Problem Set 9.3, pp. 374-375

2. $\dfrac{x^2}{16} + \dfrac{y^2}{9} = 1$

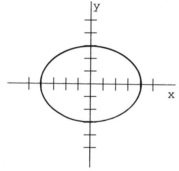

4. $\dfrac{x^2}{9} + \dfrac{y^2}{16} = 1$

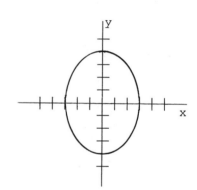

6. $\dfrac{x^2}{4} + \dfrac{y^2}{16} = 1$

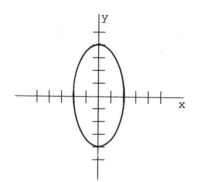

8. $\dfrac{x^2}{64} + \dfrac{y^2}{36} = 1$

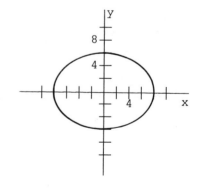

10. $\dfrac{x^2}{4} + \dfrac{y^2}{7} = 1$

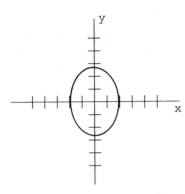

12. $4x^2 + 25y^2 = 100$

$$\dfrac{4x^2}{100} + \dfrac{25y^2}{100} = \dfrac{100}{100}$$

$$\dfrac{x^2}{25} + \dfrac{y^2}{4} = 1$$

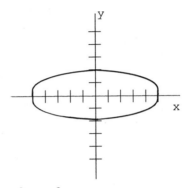

14. $x^2 + 4y^2 = 4$

$$\dfrac{x^2}{4} + \dfrac{4y^2}{4} = \dfrac{4}{4}$$

$$\dfrac{x^2}{4} + \dfrac{y^2}{1} = 1$$

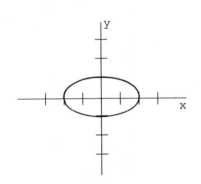

16. $x^2 + 9y^2 = 1$

$$\dfrac{x^2}{1} + \dfrac{y^2}{1/9} = 1$$

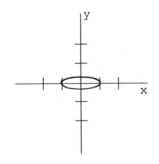

18. $\dfrac{x^2}{25} - \dfrac{y^2}{9} = 1$

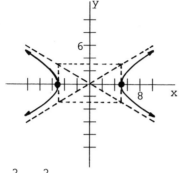

20. $\dfrac{y^2}{9} - \dfrac{x^2}{25} = 1$

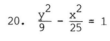

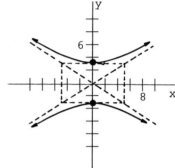

22. $\dfrac{x^2}{16} - \dfrac{y^2}{16} = 1$

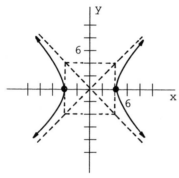

24., $\dfrac{y^2}{16} - \dfrac{x^2}{16} = 1$

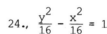

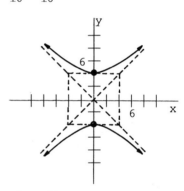

26. $4x^2 - 49y^2 = 196$

$\dfrac{4x^2}{196} - \dfrac{49y^2}{196} = \dfrac{196}{196}$

$\dfrac{x^2}{49} - \dfrac{y^2}{4} = 1$

28. $y^2 - x^2 = 1$

$\dfrac{y^2}{1} - \dfrac{x^2}{1} = 1$

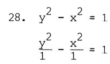

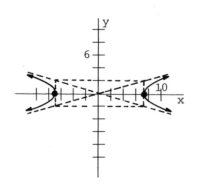

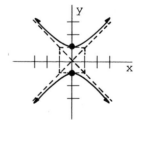

30. $\dfrac{x^2}{2^2} + \dfrac{y^2}{7^2} = 1$

   $\dfrac{x^2}{4} + \dfrac{y^2}{49} = 1$

32. $\dfrac{x^2}{2^2} - \dfrac{y^2}{6^2} = 1$

   $\dfrac{x^2}{4} - \dfrac{y^2}{36} = 1$

34. $4x^2 + 4y^2 - 64 = 0$

   $4x^2 + 4y^2 = 64$

   Divide by 4.

   $x^2 + y^2 = 16$

   A circle

36. $y^2 = 81 + x^2$

   $y^2 - x^2 = 81$

   Divide by 81.

   $\dfrac{y^2}{81} - \dfrac{x^2}{81} = 1$

   A hyperbola

38. $9x^2 = 900 - 100y^2$

   $9x^2 + 100y^2 = 900$

   Divide by 900.

   $\dfrac{x^2}{100} + \dfrac{y^2}{9} = 1$

   An ellipse

40. $y + 9x^2 = 1$

   $y = -9x^2 + 1$

   A parabola

42. $y^2 - 4x^2 = 0$

   $(y - 2x)(y + 2x) = 0$

   $y - 2x = 0$ or $y + 2x = 0$

   $y = 2x y = -2x$

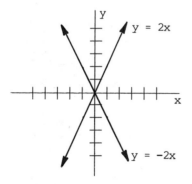

44. $2.89x^2 - 20.25y^2 = 58.5225$

   Divide by 58.5225.

   $\dfrac{x^2}{20.25} - \dfrac{y^2}{2.89} = 1$

   $\dfrac{x^2}{(4.5)^2} - \dfrac{y^2}{(1.7)^2} = 1$

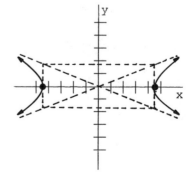

Problem Set 9.4, pp. 379-381

2. $y = x^2$

    $y = 4x - 4$

    Replace y in the second equation by x^2.

    $$x^2 = 4x - 4$$
    $$x^2 - 4x + 4 = 0$$
    $$(x - 2)^2 = 0$$
    $$x = 2$$

    Replace x in $y = x^2$ by 2.

    $$y = 2^2 = 4$$

    The solution is (2, 4).

4. $y = x^2 + 3x + 1$

    $y = x + 4$

    Replace y in the second equation by $x^2 + 3x + 1$.

    $$x^2 + 3x + 1 = x + 4$$
    $$x^2 + 2x - 3 = 0$$
    $$(x + 3)(x - 1) = 0$$
    $$x = -3 \text{ or } x = 1$$

    Substitute in $y = x + 4$.

    $$y = -3 + 4 = 1$$
    $$y = 1 + 4 = 5$$

    The solutions are (-3, 1) and (1, 5).

6. $y^2 - 4x^2 = 45$

    $y = 3x$

    Replace y in the first equation by 3x.

    $$(3x)^2 - 4x^2 = 45$$
    $$9x^2 - 4x^2 = 45$$
    $$5x^2 = 45$$
    $$x^2 = 9$$
    $$x = \pm 3$$

    Substitute in $y = 3x$.

    $$y = 3(3) = 9$$
    $$y = 3(-3) = -9$$

8. $x^2 + y^2 = 25$

    $2x + y = 5$

    Solve $2x + y = 5$ for y.

    $$y = 5 - 2x$$

    Replace y in the first equation with $5 - 2x$.

    $$x^2 + (5 - 2x)^2 = 25$$
    $$x^2 + 25 - 20x + 4x^2 = 25$$
    $$5x^2 - 20x = 0$$
    $$5x(x - 4) = 0$$
    $$5x = 0 \text{ or } x - 4 = 0$$
    $$x = 0 \qquad x = 4$$

The solutions are (3, 9) and (-3, -9).

Substitute in $y = 5 - 2x$.

$y = 5 - 2(0) = 5$

$y = 5 - 2(4) = -3$

The solutions are (0, 5) and (4, -3).

10. $x^2 + y^2 = 16$

$x - 2y = 8$

Solve the second equation for x.

$x = 2y + 8$

Substitute in the first equation.

$(2y + 8)^2 + y^2 = 16$

$4y^2 + 32y + 64 + y^2 = 16$

$5y^2 + 32y + 48 = 0$

$(5y + 12)(y + 4) = 0$

$y = -\frac{12}{5}$ or $y = -4$

Substitute in $x = 2y + 8$.

$x = 2(-\frac{12}{5}) + 8 = -\frac{16}{5}$

$x = 2(-4) + 8 = 0$

The solutions are $(-\frac{12}{5}, -\frac{16}{5})$ and (-4, 0).

14. $x^2 + y^2 = 9$

$y = x^2 - 3$

Solve $y = x^2 - 3$ for x^2.

$y + 3 = x^2$

12. $x + y = 6$

$xy = 9$

Solve $xy = 9$ for y.

$y = \frac{9}{x}$

Substitute in $x + y = 6$.

$x + \frac{9}{x} = 6$

$x^2 + 9 = 6x$

$x^2 - 6x + 9 = 0$

$(x - 3)^2 = 0$

$x = 3$

Substitute in $y = \frac{9}{x}$.

$y = \frac{9}{3} = 3$

The solution is (3, 3).

16. $x^2 + xy - y^2 = 31$

$y - x = 2$

Solve $y - x = 2$ for y.

$y = x + 2$

Substitute in $x^2 + y^2 = 9$.

$$y + 3 + y^2 = 9$$
$$y^2 + y - 6 = 0$$
$$(y + 3)(y - 2) = 0$$
$$y = -3 \qquad y = 2$$

Substitute in $x^2 + y^2 = 9$.

$$x^2 + (-3)^2 = 9 \quad x^2 + 2^2 = 9$$
$$x^2 = 0 \qquad x^2 = 5$$
$$x = 0 \qquad x = \pm\sqrt{5}$$

The solutions are $(0, -3)$, $(\sqrt{5}, 2)$, and $(-\sqrt{5}, 2)$.

Substitute in

$$x^2 + xy - y^2 = 31$$
$$x^2 + x(x + 2) - (x + 2)^2 = 31$$
$$x^2 + x^2 + 2x - x^2 - 4x - 4 = 31$$
$$x^2 - 2x - 35 = 0$$
$$(x - 7)(x + 5) = 0$$

$x = 7$ or $x = -5$

Substitute in $y = x + 2$.

$$y = 7 + 2 = 9$$
$$y = -5 + 2 = -3$$

The solutions are $(7, 9)$ and $(-5, -3)$.

18. $x^2 + y^2 = 16$

$\underline{x^2 - y^2 = 16}$

$2x^2 \qquad = 32$ Add

$$x^2 = 16$$
$$x = \pm 4$$

Substitute in $x^2 + y^2 = 16$.

$$4^2 + y^2 = 16 \qquad (-4)^2 + y^2 = 16$$
$$y^2 = 0 \qquad\qquad y^2 = 0$$
$$y = 0 \qquad\qquad y = 0$$

The solutions are $(4, 0)$ and $(-4, 0)$.

20. $3x^2 + 2y^2 = 200$

$x^2 - y^2 = 60$

Multiply $x^2 - y^2 = 60$ by 2.

$3x^2 + 2y^2 = 200$

$\underline{2x^2 - 2y^2 = 120}$

$5x^2 \qquad = 320$

$$x^2 = 64$$
$$x = \pm 8$$

Substitute in $x^2 - y^2 = 60$.

$$8^2 - y^2 = 60 \qquad (-8)^2 - y^2 = 60$$
$$-y^2 = -4 \qquad\qquad -y^2 = -4$$
$$y = \pm 2 \qquad\qquad y = \pm 2$$

The solutions are $(8, 2)$, $(8, -2)$, $(-8, 2)$, and $(-8, -2)$.

22. $x^2 + y^2 = 4$

$4x^2 + 9y^2 = 36$

Multiply $x^2 + y^2 = 4$ by -9.

$-9x^2 - 9y^2 = -36$

$\underline{4x^2 + 9y^2 = 36}$

$-5x^2 \qquad = 0$

$x^2 = 0$

$x = 0$

Substitute in $x^2 + y^2 = 4$.

$0^2 + y^2 = 4$

$y^2 = 4$

$y = \pm 2$

The solutions are $(0, 2)$ and $(0, -2)$.

24. $2x^2 - 3y^2 = 6$

$3x^2 + 2y^2 = 35$

Multiply $2x^2 - 3y^2 = 6$ by 2 and $3x^2 + 2y^2 = 35$ by 3.

$4x^2 - 6y^2 = 12$

$\underline{9x^2 + 6y^2 = 105}$

$13x^2 \qquad = 117$

$x^2 = 9$

$x = \pm 3$

Substitute in $3x^2 + 2y^2 = 35$.

$3(3)^2 + 2y^2 = 35$

$2y^2 = 8$

$y^2 = 4$

$y = \pm 2$

$3(-3)^2 + 2y^2 = 35$

$2y^2 = 8$

$y^2 = 4$

$y = \pm 2$

The solutions are $(3, 2)$, $(3, -2)$, $(-3, 2)$, and $(-3, -2)$.

26. $xy - y^2 = 30$

$3xy - y^2 = 162$

Multiply $xy - y^2 = 30$ by -3.

$-3xy + 3y^2 = -90$

$\underline{3xy - y^2 = 162}$

$2y^2 = 72$

$y^2 = 36$

$y = \pm 6$

Substitute in $xy - y^2 = 30$.

$$x(6) - 6^2 = 30 \qquad x(-6) - (-6)^2 = 30$$

$$6x = 66 \qquad\qquad -6x = 66$$

$$x = 11 \qquad\qquad x = -11$$

The solutions are $(11, 6)$ and $(-11, -6)$.

28. $9x^2 + 9y^2 = 13$

$9x^2 + 3y = 7$

Multiply $9x^2 + 3y = 7$ by -1.

$$9x^2 + 9y^2 \qquad = 13$$

$$\underline{-9x^2 \qquad - 3y = -7}$$

$$9y^2 - 3y = 6$$

$$9y^2 - 3y - 6 = 0$$

$$3y^2 - y - 2 = 0$$

$$(3y + 2)(y - 1) = 0$$

$y = -\dfrac{2}{3}$ or $y = 1$

Substitute in $9x^2 + 3y = 7$.

$$9x^2 + 3(-\tfrac{2}{3}) = 7 \qquad 9x^2 + 3(1) = 7$$

$$9x^2 = 9 \qquad\qquad 9x^2 = 4$$

$$x^2 = 1 \qquad\qquad x^2 = \frac{4}{9}$$

$$x = \pm 1 \qquad\qquad x = \pm\frac{2}{3}$$

The solutions are $(1, -\frac{2}{3})$, $(-1, -\frac{2}{3})$, $(\frac{2}{3}, 1)$, and $(-\frac{2}{3}, 1)$.

30. $x^2 - 2xy + y^2 = 1$

$\qquad x^2 + y^2 = 5$

Multiply $x^2 + y^2 = 5$ by -1.

$$x^2 - 2xy + y^2 = 1$$
$$\underline{-x^2 \qquad - y^2 = -5}$$
$$-2xy \qquad = -4$$

$$y = \frac{2}{x}$$

Substitute in $x^2 + y^2 = 5$.

$$x^2 + (\frac{2}{x})^2 = 5$$

$$x^2 + \frac{4}{x^2} = 5$$

$$x^4 + 4 = 5x^2 \qquad \text{Multiply by } x^2$$

$$x^4 - 5x^2 + 4 = 0$$

$$(x^2 - 1)(x^2 - 4) = 0$$

$$x^2 - 1 = 0 \quad \text{or} \quad x^2 - 4 = 0$$

$$x^2 = 1 \qquad\qquad x^2 = 4$$

$$x = \pm 1 \qquad\qquad x = \pm 2$$

Substitute in $y = \frac{2}{x}$.

$$y = \frac{2}{1} = 2 \qquad\qquad\qquad y = \frac{2}{2} = 1$$

$$y = \frac{2}{-1} = -2 \qquad\qquad\qquad y = \frac{2}{-2} = -1$$

The solutions are $(1, 2)$, $(-1, -2)$, $(2, 1)$, and $(-2, -1)$.

32. $x^2 + 2xy + y^2 = 25$

$\qquad x^2 + xy + y^2 = 19$

Multiply $x^2 + xy + y^2 = 19$ by -1.

$$x^2 + 2xy + y^2 = 25$$
$$\underline{-x^2 - xy - y^2 = -19}$$
$$xy \qquad = 6$$

$$y = \frac{6}{x}$$

Substitute in $x^2 + xy + y^2 = 19$.

$$x^2 + x(\frac{6}{x}) + (\frac{6}{x})^2 = 19$$

$$x^2 + 6 + \frac{36}{x^2} = 19$$

$$x^2 - 13 + \frac{36}{x^2} = 0$$

$$x^4 - 13x^2 + 36 = 0 \quad \text{Multiply by } x^2$$

$$(x^2 - 4)(x^2 - 9) = 0$$

$$x^2 = 4 \quad \text{or} \quad x^2 = 9$$

$$x = \pm 2 \qquad x = \pm 3$$

Substitute in $y = \frac{6}{x}$.

$$y = \frac{6}{2} = 3 \qquad\qquad\qquad y = \frac{6}{3} = 2$$

$$y = \frac{6}{-2} = -3 \qquad\qquad\qquad y = \frac{6}{-3} = -2$$

The solutions are $(2, 3)$, $(-2, -3)$, $(3, 2)$, and $(-3, -2)$.

34. $11x^2 - y^2 = 45$

$$y = 4x$$

Replace y by $4x$ in the first equation.

$$11x^2 - (4x)^2 = 45$$

$$11x^2 - 16x^2 = 45$$

$$-5x^2 = 45$$

$$x^2 = -9$$

$$x = \pm\sqrt{-9}$$

$$x = \pm 3i$$

Substitute in $y = 4x$.

$$y = 4(3i) = 12i$$

$$y = 4(-3i) = -12i$$

36. $2x^2 + 3y^2 = 19$

$$x^2 + y^2 = 5$$

Multiply $x^2 + y^2 = 5$ by -3.

$$2x^2 + 3y^2 = 19$$

$$\underline{-3x^2 - 3y^2 = -15}$$

$$-x^2 \qquad\quad = 4$$

$$x^2 = -4$$

$$x = \pm\sqrt{-4}$$

$$x = \pm 2i$$

Substitute in $x^2 + y^2 = 5$.

$$(2i)^2 + y^2 = 5 \qquad (-2i)^2 + y^2 = 5$$

$$-4 + y^2 = 5 \qquad\qquad -4 + y^2 = 5$$

$$y^2 = 9 \qquad\qquad\qquad y^2 = 9$$

$$y = \pm 3 \qquad\qquad\qquad y = \pm 3$$

The solutions are (3i, 12i) and (-3i, -12i).

The solutions are (2i, 3), (2i, -3), (-2i, 3), and (-2i, -3).

38. $x^2 + 2xy - y^2 = 7$

$x^2 - y^2 = 3$

Multiply $x^2 - y^2 = 3$ by -1.

$x^2 + 2xy - y^2 = 7$

$\underline{-x^2 \qquad + y^2 = -3}$

$\qquad 2xy \qquad = 4$

$\qquad y = \dfrac{2}{x}$

Substitute in $x^2 - y^2 = 3$

$x^2 - (\dfrac{2}{x})^2 = 3$

$x^2 - \dfrac{4}{x^2} = 3$

$x^4 - 4 = 3x^2$

$x^4 - 3x^2 - 4 = 0$

$(x^2 - 4)(x^2 + 1) = 0$

$x^2 = 4$ or $x^2 = -1$

$x = \pm 2 \qquad x = \pm\sqrt{-1} = \pm i$

Substitute in $y = \dfrac{2}{x}$.

$y = \dfrac{2}{2} = 1$

$y = \dfrac{2}{-2} = -1$

40. $\dfrac{1}{2}xy = 104$

$x(y + 3) = 247$

Simplify the system.

$xy = 208$

$xy + 3x = 247$

Replace xy by 208 in $xy + 3x = 247$.

$208 + 3x = 247$

$3x = 39$

$x = 13$

Substitute in xy = 208.

$13y = 208$

$y = 16$

$y = \dfrac{2}{i} = \dfrac{2}{i}\cdot\dfrac{-i}{-i} = \dfrac{-2i}{-i^2} = -2i$

$y = \dfrac{2}{-i} = \dfrac{2}{-i}\cdot\dfrac{i}{i} = \dfrac{2i}{-i^2} = 2i$

The solutions are (2, 1), (-2, -1), (i, -2i), and (-i, 2i).

42.   x = width of rectangle

y = length of rectangle

$$xy = 12$$
$$x^2 + y^2 = 5^2$$

Solve xy = 12 for y.

$$y = \frac{12}{x}$$

Substitute in $x^2 + y^2 = 5^2$.

$$x^2 + (\tfrac{12}{x})^2 = 5^2$$
$$x^2 + \frac{144}{x^2} = 25$$
$$x^4 + 144 = 25x^2 \qquad \text{Multiply by } x^2$$
$$x^4 - 25x^2 + 144 = 0$$
$$(x^2 - 9)(x^2 - 16) = 0$$
$$x^2 = 9 \quad \text{or} \quad x^2 = 16$$
$$x = \pm 3 \qquad\qquad x = \pm 4$$

Substitute the positive x-values in $y = \frac{12}{x}$.

$$y = \frac{12}{3} = 4 \qquad\qquad\qquad y = \frac{12}{4} = 3$$

The width is 3 feet; the length is 4 feet.

44. $5.41x^2 + 6.84y^2 = 936.67$

$$x^2 - y^2 = 1.75$$

Multiply $x^2 - y^2 = 1.75$ by 6.84.

$$5.41x^2 + 6.84y^2 = 936.67$$
$$\underline{6.84x^2 - 6.84y^2 = 11.97}$$
$$12.25x^2 = 948.64$$
$$x^2 = 77.44$$
$$x = \pm 8.8$$

Substitute in $x^2 - y^2 = 1.75$.

$(8.8)^2 - y^2 = 1.75$ $\qquad\qquad$ $(-8.8)^2 - y^2 = 1.75$

$\qquad -y^2 = -75.69$ $\qquad\qquad\qquad -y^2 = -75.69$

$\qquad y^2 = 75.69$ $\qquad\qquad\qquad\qquad y^2 = 75.69$

$\qquad y = \pm 8.7$ $\qquad\qquad\qquad\qquad y = \pm 8.7$

The solutions are $(8.8, 8.7)$, $(8.8, -8.7)$, $(-8.8, 8.7)$, and $(-8.8, -8.7)$.

Problem Set 9.5, pp. 384-385

2.

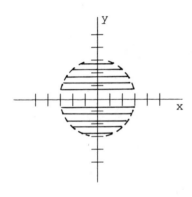

4.

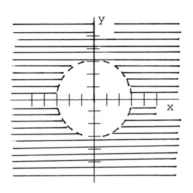

6.

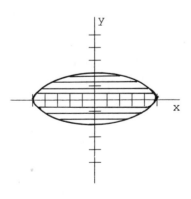

8.

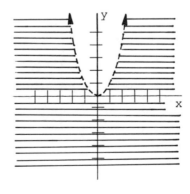

10.

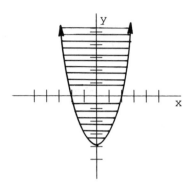

12.

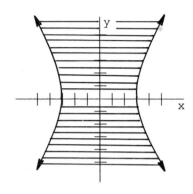

14.

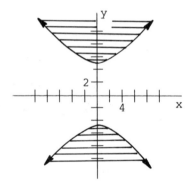

16.

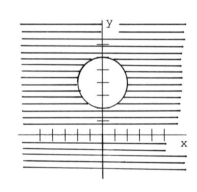

18.

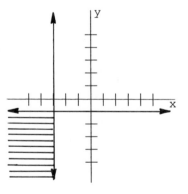

20.

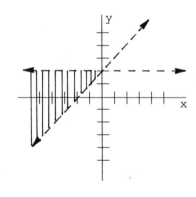

22.

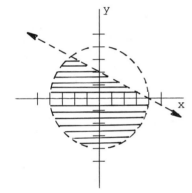

24.

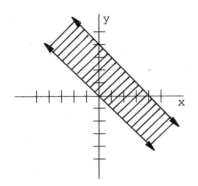

26.

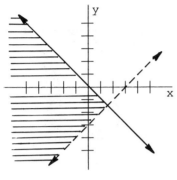

28.

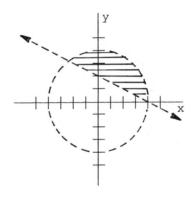

30.

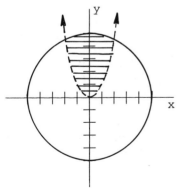

32.

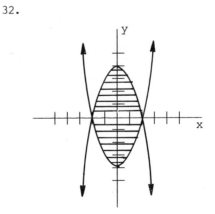

34.

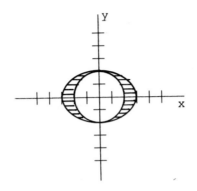

36.

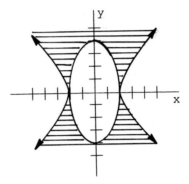

38.

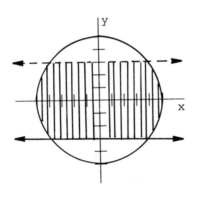

40. $1.4x^2 + 1.4y^2 < 8.75$

$1.96x^2 + y^2 > 12.25$

Divide the first equation by
1.4 and the second by 12.25.

$$x^2 + y^2 < 6.25$$

$$\frac{x^2}{6.25} + \frac{y^2}{12.25} > 1$$

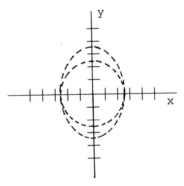

Since there are no points
both inside the circle and
outside the ellipse, there
is no graph.

CHAPTER 10

FUNCTIONS

Problem Set 10.1, pp. 394-397

2. Domain = $\left\{8,\ -5,\ 10,\ 0\right\}$

    Range = $\left\{1,\ 7,\ 2\right\}$

    A function

4. Domain = $\left\{6,\ 4\right\}$

    Range = $\left\{9,\ 8,\ -3\right\}$

    Not a function because 4
    corresponds to both 8 and -3.

6. Domain = $\left\{0,\ -1,\ -2\right\}$

    Range = $\left\{0,\ -1,\ -2\right\}$

    A function

8. Domain = $\left\{8,\ 9,\ 10\right\}$

    Range = $\left\{11\right\}$

    A function

10. Domain = $\left\{1,\ 2,\ 3\right\}$

    Range = $\left\{2,\ 3,\ 4\right\}$

    A function

12. Yes

14. Yes

16. Yes

18. No. When x = 5, y = ±2.

20. Yes

22. No. When x = 0, y = ±4.

24. Yes

26. No. When x = 3, y can be any number greater than 2.

28. All real numbers

30. All real numbers

32. x ≠ 2

34. x ≠ 5, -5

36. All real numbers

38. $y = \dfrac{x}{x^2 - 7x + 12}$

$$y = \frac{x}{(x - 3)(x - 4)}$$

$$x \neq 3, 4$$

40. $x - 7 \geq 0$

$x \geq 7$

42. All real numbers

44. $3x + 8 \geq 0$

$3x \geq -8$

$x \geq -\dfrac{8}{3}$

46. $24 - 4x > 0$

$-4x > -24$

$x < 6$

48. Domain: $-2 \leq x \leq 2$

Range: $-5 \leq y \leq 5$

Not a function

50. Domain: All reals

Range: $y \geq 3$

A function

52. Domain: All reals

Range: All reals

A function

54. Domain: $x = 2$

Range: All reals

Not a function

56. Domain: All reals

Range: $y \leq -1$ or $y \geq 1$

Not a function

58. Domain: All reals

Range: All reals

A function

60. Domain: $-3 \leq x \leq 3$

Range: $-3 \leq y \leq 0$

A function

62. $P = 4x$

Domain: $x > 0$

64. $y = 250 + 0.15x$

66. $p = 600 - 3q$

Domain: $0 \leq q \leq 200$

68. $1.550x - 99.014 \geq 0$

$1.550x \geq 99.014$

$x \geq 63.88$

Problem Set 10.2, pp. 401–402

2. $f(4) = 2(4) + 1 = 9$

4. $f(-4) = 2(-4) + 1 = -7$

6. $f(1) = 2(1) + 1 = 3$

8. $f(\frac{5}{2}) = 2(\frac{5}{2}) + 1 = 6$

10. $f(b) = 2b + 1$

12. $f(b - 1) = 2(b - 1) + 1$
$$= 2b - 2 + 1$$
$$= 2b - 1$$

14. $f(3) = 6(3)^2 - 4(3) + 3 = 54 - 12 + 3 = 45$

16. $f(-3) = 6(-3)^2 - 4(-3) + 3 = 54 + 12 + 3 = 69$

18. $f(\frac{2}{3}) = 6(\frac{2}{3})^2 - 4(\frac{2}{3}) + 3 = \frac{8}{3} - \frac{8}{3} + \frac{27}{3} = \frac{27}{3}$

20. $f(0) = 6(0)^2 - 4(0) + 3 = 0 - 0 + 3 = 3$

22. $f(-b) = 6(-b)^2 - 4(-b) + 3 = 6b^2 + 4b + 3$

24. $f(x - h) = 6(x - h)^2 - 4(x - h) + 3$
$$= 6(x^2 - 2xh + h^2) - 4x + 4h + 3$$
$$= 6x^2 - 12xh + 6h^2 - 4x + 4h + 3$$

26. $g(2) = 8$

28. $g(-8) = 8$

30. $g(\frac{1}{2}) = 8$

32. $g(b) = 8$

34. $f(4) + g(4) = [3(4) + 2] + [4^2 - 5] = 14 + 11 = 25$

36. $f(\frac{2}{3}) \cdot g(-2) = [3(\frac{2}{3}) + 2][(-2)^2 - 5] = [4][-1] = -4$

38. $\dfrac{f(6)}{g(0)} = \dfrac{3(6) + 2}{0^2 - 5} = \dfrac{20}{-5} = -4$

40. $f(g(1)) = f(1^2 - 5) = f(-4) = 3(-4) + 2 = -10$

42. $g(f(1)) = g(3(1) + 2) = g(5) = 5^2 - 5 = 20$

44. $f(2) - g(8) = 6 - (-8) = 14$

46. $f(0) \cdot g(6) = (-4) \cdot 7 = -28$

48. $g(f(-7)) = g(4) = 2$

50. $g(g(8)) = g(-8) = 2$

52. $f(g(f(2))) = f(g(6)) = f(7) = 9$

54. $\dfrac{f(t) - f(a)}{t - a}$

$= \dfrac{5t + 1 - (5a + 1)}{t - a}$

$= \dfrac{5t - 5a}{t - a}$

$= \dfrac{5(t - a)}{t - a}$

$= 5$

56. $\dfrac{f(t) - f(a)}{t - a}$

$= \dfrac{t^2 + 5 - (a^2 + 5)}{t - a}$

$= \dfrac{t^2 - a^2}{t - a}$

$= \dfrac{(t + a)(t - a)}{t - a}$

$= t + a$

58. $\dfrac{f(t) - f(a)}{t - a}$

$= \dfrac{t^3 + 4 - (a^3 + 4)}{t - a}$

$= \dfrac{t^3 - a^3}{t - a}$

$= \dfrac{(t - a)(t^2 + ta + a^2)}{t - a}$

$= t^2 + ta + a^2$

60. $\dfrac{f(t) - f(a)}{t - a}$

$= \dfrac{7 - 7}{t - a}$

$= \dfrac{0}{t - a}$

$= 0$

62. $f(g(a)) = f(3a) = (3a)^2 - 1 = 9a^2 - 1$

$\quad g(f(a)) = g(a^2 - 1) = 3(a^2 - 1) = 3a^2 - 3$

64. C(x) = 8x + 350

C(0) = 8(0) + 350 = $350 is the cost of producing 0 motors.

C(125) = 8(125) + 350 = $1350 is the cost of producing 125 motors.

66. $V(r) = 8500(1 + r)^{10}$

$V(0.06) = 8500(1 + 0.06)^{10} \approx \$15,222.21$ is the value when the interest rate is 6%.

$V(0.1375) = 8500(1 + 0.1375)^{10} \approx \$30,827.12$ is the value when the interest rate is 13.75%.

Problem Set 10.3, pp. 406 - 407

2.

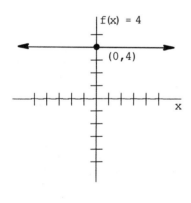

4.

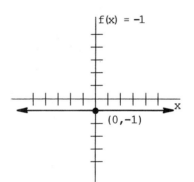

6.

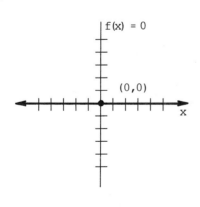

8.

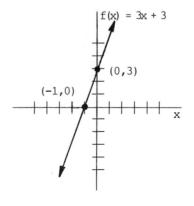

10.

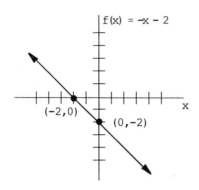

12.

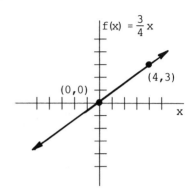

14.

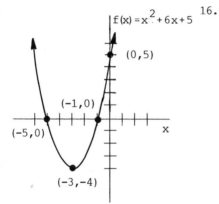

16.

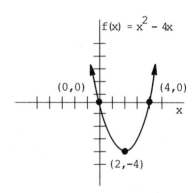

18.

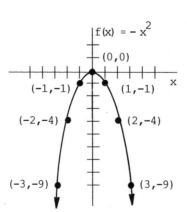

20. Linear function

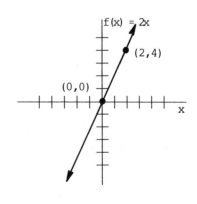

22. Quadratic function

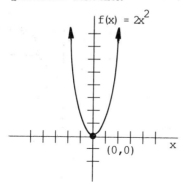

$f(x) = 2x^2$

$(0,0)$

24. Constant function

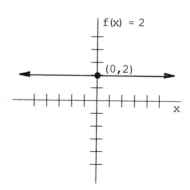

$f(x) = 2$

$(0,2)$

26. Quadratic function

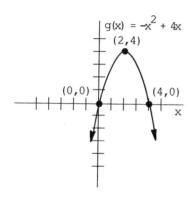

$g(x) = -x^2 + 4x$

$(2,4)$

$(0,0)$ $(4,0)$

28. Linear function

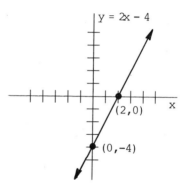

$y = 2x - 4$

$(2,0)$

$(0,-4)$

30. Quadratic function

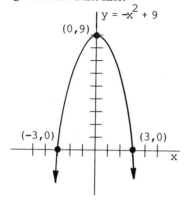

$y = -x^2 + 9$

$(0,9)$

$(-3,0)$ $(3,0)$

32. Constant function

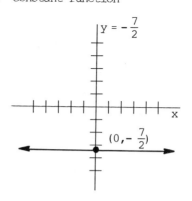

$y = -\dfrac{7}{2}$

$(0,-\dfrac{7}{2})$

34. Quadratic function

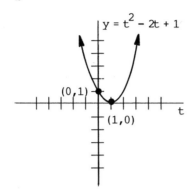

$y = t^2 - 2t + 1$

(0,1)

(1,0)

t

36. Quadratic function

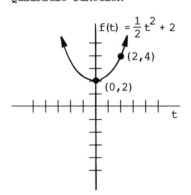

$f(t) = \frac{1}{2}t^2 + 2$

(2,4)

(0,2)

t

38. $f(4) = 0$

40. $f(5) = -2$

42. $f(-3)$ is undefined.

44.

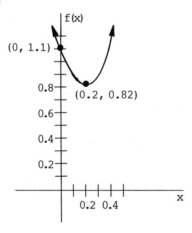

f(x)

(0, 1.1)

0.8

(0.2, 0.82)

0.6

0.4

0.2

0.2 0.4

x

Problem Set 10.4, p. 412

2.

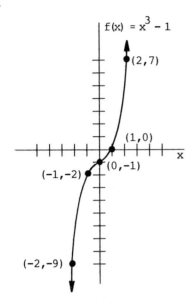

4.

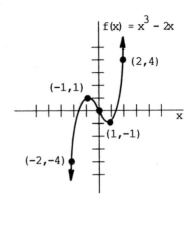

6.

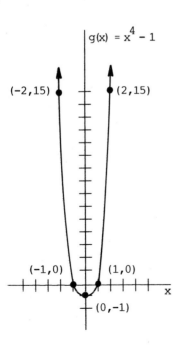

8.

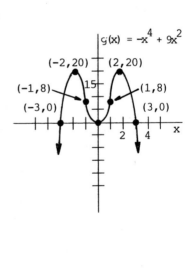

10.

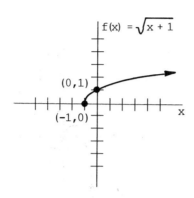

12.

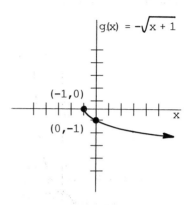

14.

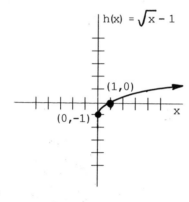

16.

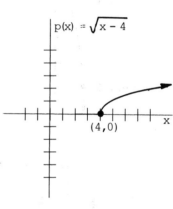

18.

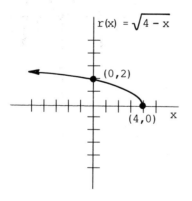

20.

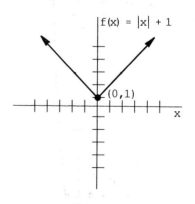

22.

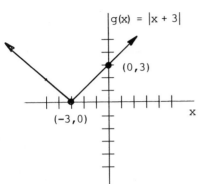

24.

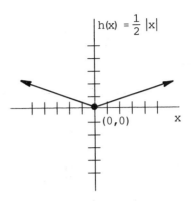

26.

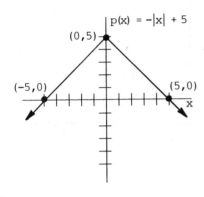

28.

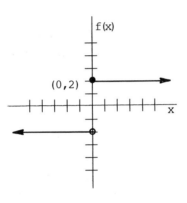

30.

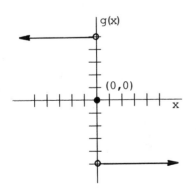

32.

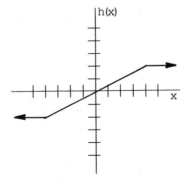

34.   36. 2.6 hr < t < 6.8 hr

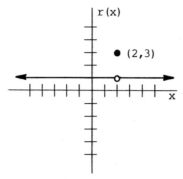

38. $$C(x) = \begin{cases} 5 \text{ if } 0 < x \le 1 \\ 8 \text{ if } 1 < x \le 2 \\ 11 \text{ if } 2 < x \le 3 \\ 14 \text{ if } 3 < x \le 4 \end{cases}$$

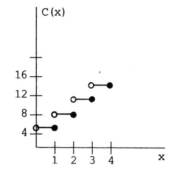

40.

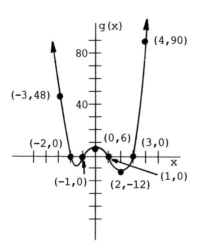

Problem Set 10.5, pp. 416-418

2. $f^{-1} = \left\{(4, 3), (3, -2), (0, -1)\right\}$

f^{-1} is a function.

4. $f^{-1} = \left\{(3, 3), (-1, -1), (5, -5)\right\}$

f^{-1} is a function.

6. $f^{-1} = \left\{(-3, 0), (7, 2), (-1, -5), (7, -4)\right\}$

f^{-1} is not a function because x = 7 corresponds to y = 2 and
y = -4.

8. f(x) = x + 4

    y = x + 4

    Equation for f^{-1} is

        x = y + 4

        y = x - 4

    $f^{-1}(x) = x - 4.$

    f^{-1} is a function

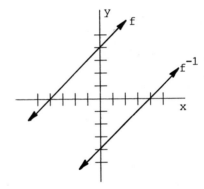

10. f(x) = 3x

    y = 3x

    Equation for f^{-1} is

        x = 3y

        $y = \frac{1}{3}x$

    $f^{-1}(x) = \frac{1}{3}x.$

    f^{-1} is a function.

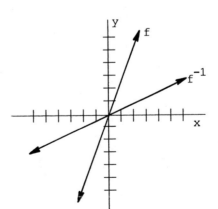

12. $f(x) = 2x - 6$

$\quad\quad y = 2x - 6$

Equation for f^{-1} is

$\quad\quad x = 2y - 6$

$\quad\quad 2y = x + 6$

$\quad\quad y = \frac{1}{2}x + 3$

$f^{-1}(x) = \frac{1}{2}x + 3.$

f^{-1} is a function.

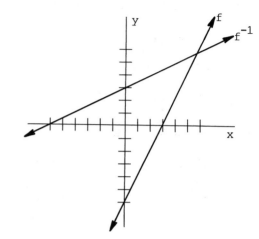

14. $f(x) = -x + 5$

   $y = -x + 5$

   Equation for f^{-1} is

   $x = -y + 5$

   $y = -x + 5$

   $f^{-1}(x) = -x + 5.$

   f^{-1} is a function.

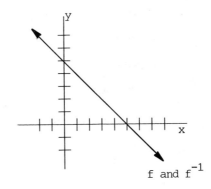

f and f^{-1}

16. $f(x) = x^2 + 3$

   $y = x^2 + 3$

   Equation for f^{-1} is

   $x = y^2 + 3$

   $y^2 = x - 3$

   $y = \pm\sqrt{x - 3}$

   $f^{-1}(x) = \pm\sqrt{x - 3}.$

   f^{-1} is not a function.

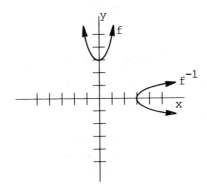

18. $f(x) = x^2 + 3,\ x \geq 0$

   $y = x^2 + 3,\ x \geq 0$

   Equation for f^{-1} is

   $x = y^2 + 3,\ y \geq 0$

   $y^2 = x - 3,\ y \geq 0$

   $y = \pm\sqrt{x - 3},\ y \geq 0$

   $f^{-1}(x) = \sqrt{x - 3}.$

   f^{-1} is a function.

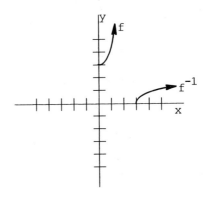

20. $f(x) = 4$

$y = 4$

Equation for f^{-1} is

$x = 4$.

f^{-1} is not a function.

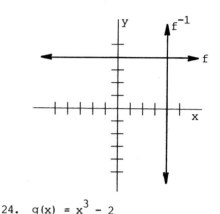

22. $f(x) = 3x^2 - 9$

$y = 3x^2 - 9$

Equation for f^{-1} is

$x = 3y^2 - 9$

$3y^2 = x + 9$

$y^2 = \dfrac{x + 9}{3}$

$y = \pm\sqrt{\dfrac{x + 9}{3}}$

$f^{-1}(x) = \pm\sqrt{\dfrac{x + 9}{3}}$.

24. $g(x) = x^3 - 2$

$y = x^3 - 2$

Equation for g^{-1} is

$x = y^3 - 2$

$y^3 = x + 2$

$y = \sqrt[3]{x + 2}$

$g^{-1}(x) = \sqrt[3]{x + 2}$.

26. $h(x) = -x$

$y = -x$

Equation for h^{-1} is

$x = -y$

$y = -x$

$h^{-1}(x) = -x$.

28. $2x - 5y = 10$

Equation for the inverse function is

$2y - 5x = 10$.

30.   a) $f(3) = 7$

b) $f^{-1}(7) = 3$

c) $f^{-1}(f(-4)) = f^{-1}(6) = -4$ d) $f(f^{-1}(8)) = f(-1) = 8$

32. $f(x) = x + 1$

   $y = x + 1$

   Equation for f^{-1} is

   $x = y + 1$

   $y = x - 1$

   $f^{-1}(x) = x - 1.$

   a) $f(7) = 7 + 1 = 8$

   b) $f^{-1}(8) = 8 - 1 = 7$

   c) $f^{-1}(f(a)) = f^{-1}(a + 1) = (a + 1) - 1 = a$

   d) $f(f^{-1}(a)) = f(a - 1) = (a - 1) + 1 = a$

34.   36.

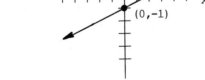

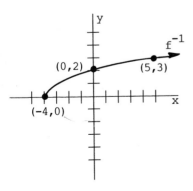

38.   40.

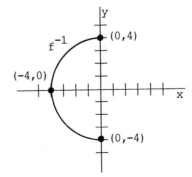

42.

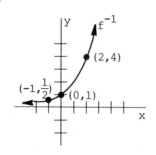

44. $f(x) = x^3$

$y = x^3$

Equation for f^{-1} is

$x = y^3$

$y = \sqrt[3]{x}$

$f^{-1}(x) = \sqrt[3]{x}.$

a) $f(19) = 19^3 = 6859$

b) $f^{-1}(6859) = \sqrt[3]{6859} = 19$

c) $f^{-1}(9261) = \sqrt[3]{9261} = 21$

d) $f^{-1}(f(1.4)) = f^{-1}(1.4^3)$

$= f^{-1}(2.744)$

$= \sqrt[3]{2.744}$

$= 1.4$

Problem Set 10.6, pp. 422-424

2. $r = ks$

4. $p = kq^2$

6. $z = \dfrac{k}{w}$

8. $v = \dfrac{k}{u^3}$

10. $y = kxz^2$

12. $T = \dfrac{kr}{s^3}$

14. $N = \dfrac{kp_1p_2}{d}$

16.   a) $y = kx$

$40 = k(5)$

$k = 8$

18.   a) $y = kx^2$

$18 = k(3)^2$

$k = 2$

b) $y = 8x$

c) $y = 8(6) = 48$

b) $y = 2x^2$

c) $y = 2(6)^2 = 72$

20. a) $y = \dfrac{k}{x}$

$7 = \dfrac{k}{4}$

$k = 28$

b) $y = \dfrac{28}{x}$

c) $y = \dfrac{28}{6} = \dfrac{14}{3}$

22. $y = kxz$

$16 = k(4)(3)$

$k = \dfrac{16}{12} = \dfrac{4}{3}$

$y = \dfrac{4}{3}xz$

$y = \dfrac{4}{3}(6)(5) = 40$

24. $y = kxz^2$

$96 = k(\dfrac{3}{4})(8)^2$

$96 = k(48)$

$k = 2$

$y = 2xz^2$

$y = 2(6)(\dfrac{1}{2})^2 = 3$

26. $y = kx$

$4 = k(100)$

$k = \dfrac{4}{100} = \dfrac{1}{25}$

$y = \dfrac{1}{25}x$

$y = \dfrac{1}{25}(175) = 7$ cm

28. $p = kv^2$

$0.3 = k(10)^2$

$k = \dfrac{0.3}{100} = 0.003$

$p = 0.003v^2$

$p = 0.003(100)^2$

$\quad = 0.003(10,000)$

$\quad = 30$ lb/ft^2

Area $= 20 \cdot 40 = 800$ ft^2

Total force $= 30(800) = 24,000$ lb

30. $V = \dfrac{k}{P}$

$100 = \dfrac{k}{15}$

$k = 1500$

$V = \dfrac{1500}{P}$

$V = \dfrac{1500}{10}$

$\quad = 150$ in.^{3}

32. $R = \dfrac{k}{d^2}$

   $2 = \dfrac{k}{(0.03)^2}$

   $2 = \dfrac{k}{0.0009}$

   $k = 0.0018$

   $R = \dfrac{0.0018}{d^2}$

   $R = \dfrac{0.0018}{(0.01)^2}$

   $ = \dfrac{0.0018}{0.0001}$

   $ = 18 \text{ ohm}$

36. $I = \dfrac{k}{d^2}$

   $I = \dfrac{k}{10^2} = \dfrac{k}{100}$

   $I = \dfrac{k}{20^2} = \dfrac{k}{400}$

   Illumination from 10 ft is
   4 times greater than from
   20 ft.

34. $L = kwd^2$

   $600 = k(4)(5)^2$

   $600 = k(100)$

   $k = 6$

   $L = 6wd^2$

   $L = 6(3)(12)^2$

   $ = 2592 \text{ lb}$

38. $y = kxw^2$

   $32{,}076 = k(16.2)(15)^2$

   $k = 8.8$

   $y = 8.8xw^2$

   $y = 8.8(23.4)(25)^2$

   $ = 128{,}700$

CHAPTER 11

EXPONENTIAL AND LOGARITHMIC FUNCTIONS

Problem Set 11.1, pp. 431-432

2.

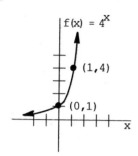

4.

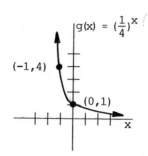

6.

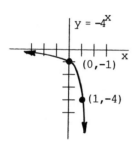

8.   Same as Problem 4

10.

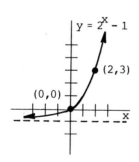

12.

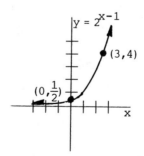

14.

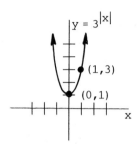

16.

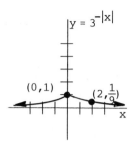

18.

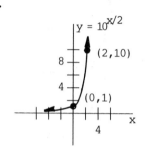

20. $V = 2(3)^t$

a) $V = 2(3)^0 = \$2$

b) $V = 2(3)^1 = \$6$

c) $V = 2(3)^2 = \$18$

d) $V = 2(3)^3 = \$54$

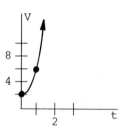

22. $Q = 100(2)^{-t/1600}$

a) $Q = 100(2)^{-0/1600} = 100(2)^0 = 100(1) = 100$ g

b) $Q = 100(2)^{-1600/1600} = 100(2)^{-1} = 100(\frac{1}{2}) = 50$ g

c) $Q = 100(2)^{-3200/1600} = 100(2)^{-2} = 100(\frac{1}{4}) = 25$ g

d) $Q = 100(2)^{-4800/1600} = 100(2)^{-3} = 100(\frac{1}{8}) = 12.5$ g

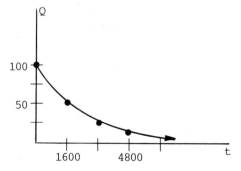

24. $PV = p[\dfrac{1 - (1 + r/12)^{-12t}}{r/12}]$

If t = 30 years, then

$40,000 = p[\dfrac{1 - (1 + 0.12/12)^{-12(30)}}{0.12/12}]$

$40,000 = p[\dfrac{1 - (1.01)^{-360}}{0.01}]$

$40,000 \approx p[\dfrac{1 - 0.0278166892}{.01}]$

$40,000 \approx p[97.21833108]$

$\qquad p \approx \$411.45.$

Total amount repaid $\approx 411.45(360) \approx \$148,120.$

If t = 40 years, then

$40,000 = p[\dfrac{1 - (1 + 0.12/12)^{-12(40)}}{0.01}]$

$40,000 = p[\dfrac{1 - (1.01)^{-480}}{0.01}]$

$40,000 \approx p[\dfrac{1 - 0.0084283116}{0.01}]$

$40,000 \approx p[99.15716884]$

$\qquad p \approx \$403.40.$

Total amount repaid $\approx 403.40(480) \approx \$193,630.$

Problem Set 11.2, pp. 436-438

2. $\log_7 49 = 2$

4. $\log_2 32 = 5$

6. $\log_2 \frac{1}{2} = -1$

8. $\log_{10} 1000 = 3$

10. $\log_{10} 0.0001 = -4$

12. $\log_8 8 = 1$

14. $\log_{100} 10 = \frac{1}{2}$

16. $\log_b v = u$

18. $3^3 = 27$

20. $5^4 = 625$

22. $(\frac{1}{9})^{-2} = 81$

24. $36^{1/2} = 6$

26. $(\frac{1}{3})^3 = \frac{1}{27}$

28. $(\sqrt{5})^2 = 5$

30. $12^0 = 1$

32. $b^S = N$

34. $4^y = 64$

    $4^y = 4^3$

    $y = 3$

36. $4^y = \frac{1}{64}$

    $4^y = 4^{-3}$

    $y = -3$

38. $10^y = 100,000$

    $10^y = 10^5$

    $y = 5$

40. $10^y = 0.001$

    $10^y = 10^{-3}$

    $y = -3$

42. $(\frac{1}{3})^y = 9$

    $(3^{-1})^y = 3^2$

    $3^{-y} = 3^2$

    $-y = 2$

    $y = -2$

44. $5^y = 1$

    $5^y = 5^0$

    $y = 0$

46. $5^y = 5$

    $5^y = 5^1$

    $y = 1$

48. $7^y = \sqrt{7}$

    $7^y = 7^{1/2}$

    $y = \frac{1}{2}$

50. $b^y = b^t$

    $y = t$

52. $x = 4^2$

    $x = 16$

54. $x = 27^{2/3}$

    $x = (\sqrt[3]{27})^2$

    $x = 9$

56. $x = 3^{-4}$

    $x = \frac{1}{3^4}$

    $x = \frac{1}{81}$

58. $a^2 = 64$

    $a = 8$ or $a \cancel{=} -8$

60. $a^3 = \frac{1}{125}$

    $a = \frac{1}{5}$

62. $a^{1/2} = 9$

$(a^{1/2})^2 = 9^2$

$a = 81$

64. $y = \log_4 x$

$x = 4^y$

x	y
1/16	-2
1/4	-1
1	0
4	1
16	2

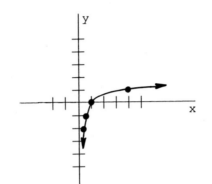

66. $y = \log_{1/4} x$

$x = (\frac{1}{4})^y$

x	y
1/16	2
1/4	1
1	0
4	-1
16	-2

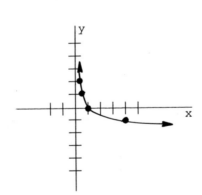

68. If $y = \log_2 (-4)$, then $2^y = -4$. But 2^y is positive for all real values of y.

70. $L = 10 \log_{10} \dfrac{I}{I_0}$

a) $L = 10 \log_{10} \dfrac{I_0}{I_0} = 10 \log_{10} 1 = 10(0) = 0$ db

b) $L = 10 \log_{10} \dfrac{10^{6.5} I_0}{I_0} = 10 \log_{10} 10^{6.5} = 10(6.5) = 65$ db

c) $L = 10 \log_{10} \dfrac{10^{12} I_0}{I_0} = 10 \log_{10} 10^{12} = 10(12) = 120$ db

72. $\log_{10} 50 \approx 1.6990$ 74. $\log_{10} 2010 \approx 3.3032$

Problem Set 11.3, pp. 441-443

2. $\log_3 2 \cdot 5 = \log_3 2 + \log_3 5$

4. $\log_5 xy = \log_5 x + \log_5 y$

6. $\log_7 9x = \log_7 9 + \log_7 x$

8. $\log_{10} x(x + 3) = \log_{10} x + \log_{10} (x + 3)$

10. $\log_2 \dfrac{5}{7} = \log_2 5 - \log_2 7$

12. $\log_4 \dfrac{x}{y} = \log_4 x - \log_4 y$

14. $\log_5 \dfrac{6}{x} = \log_5 6 - \log_5 x$

16. $\log_{10} \dfrac{x - 9}{x} = \log_{10} (x - 9) - \log_{10} x$

18. $\log_6 x^4 = 4 \log_6 x$

20. $\log_{10} x^{-2} = -2 \log_{10} x$

22. $\log_2 5^{1/2} = \dfrac{1}{2} \log_2 5$

24. $\log_a \sqrt[4]{x} = \log_a x^{1/4} = \dfrac{1}{4} \log_a x$

26. $\log_{10} 7 \cdot 100 = \log_{10} 7 + \log_{10} 100$

$= \log_{10} 7 + \log_{10} 10^2$

$= \log_{10} 7 + 2$

28. $\log_4 64x = \log_4 64 + \log_4 x$

$\qquad = \log_4 4^3 + \log_4 x$

$\qquad = 3 + \log_4 x$

30. $\log_3 \dfrac{81}{2} = \log_3 81 - \log_3 2$

$\qquad = \log_3 3^4 - \log_3 2$

$\qquad = 4 - \log_3 2$

32. $\log_2 \dfrac{1}{x} = \log_2 1 - \log_2 x = 0 - \log_2 x = -\log_2 x$

34. $\log_2 x^5 y = \log_2 x^5 + \log_2 y = 5 \log_2 x + \log_2 y$

36. $\log_2 16x^2 = \log_2 16 + \log_2 x^2 = \log_2 2^4 + 2 \log_2 x$

$\qquad = 4 + 2 \log_2 x$

38. $\log_3 \dfrac{x^3}{y} = \log_3 x^3 - \log_3 y = 3 \log_3 x - \log_3 y$

40. $\log_3 \dfrac{x^4}{27} = \log_3 x^4 - \log_3 27 = 4 \log_3 x - \log_3 3^3$

$\qquad = 4\log_3 x - 3$

42. $\log_5 \dfrac{x\sqrt{y}}{z^4} = \log_5 x\sqrt{y} - \log_5 z^4$

$\qquad = \log_5 x + \log_5 y^{1/2} - 4 \log_5 z$

$\qquad = \log_5 x + \dfrac{1}{2} \log_5 y - 4 \log_5 z$

44. $\log_4 \dfrac{x^2}{yz} = \log_4 x^2 - \log_4 yz$

$\qquad = 2 \log_4 x - (\log_4 y + \log_4 z)$

$\qquad = 2 \log_4 x - \log_4 y - \log_4 z$

46. $\log_4 \dfrac{x}{4y} = \log_4 x - \log_4 4y$

$\qquad = \log_4 x - (\log_4 4 + \log_4 y)$

$\qquad = \log_4 x - (1 + \log_4 y)$

$\qquad = \log_4 x - 1 - \log_4 y$

48. $8 \log_5 x = \log_5 x^8$

50. $\frac{1}{3} \log_2 x = \log_2 x^{1/3} = \log_2 \sqrt[3]{x}$

52. $-\log_{10} x = (-1) \cdot \log_{10} x = \log_{10} x^{-1} = \log_{10} \frac{1}{x}$

54. $\log_3 r + \log_3 s = \log_3 rs$

56. $\log_2 r - \log_2 s = \log_2 \frac{r}{s}$

58. $\log_7 x + \log_7 y + \log_7 z = \log_7 xyz$

60. $\log_{10} x - \log_{10} y + \log_{10} z = (\log_{10} x + \log_{10} z) - \log_{10} y$

$$= \log_{10} xz - \log_{10} y$$

$$= \log_{10} \frac{xz}{y}$$

62. $\frac{1}{4} \log_a x - \log_a y - \log_a z = \log_a x^{1/4} - (\log_a y + \log_a z)$

$$= \log_a \sqrt[4]{x} - \log_a yz$$

$$= \log_a \frac{\sqrt[4]{x}}{yz}$$

64. $\log_a x - \log_a y - 3 \log_a z = \log_a x - \log_a y - \log_a z^3$

$$= \log_a x - (\log_a y + \log_a z^3)$$

$$= \log_a x - \log_a yz^3$$

$$= \log_a \frac{x}{yz^3}$$

66. $\log_{10} (x + 2) + \log_{10} (x - 2) = \log_{10} (x + 2)(x - 2)$

$$= \log_{10} (x^2 - 4)$$

68. $\log_{10} (x^2 + 8x + 15) - \log_{10} (x + 5) = \log_{10} \frac{x^2 + 8x + 15}{x + 5}$

$$= \log_{10} \frac{(x + 3)(x + 5)}{x + 5}$$

$$= \log_{10} (x + 3)$$

70. $\log_2 (4 + 4) = \log_2 8 = \log_2 2^3 = 3$

    $\log_2 4 + \log_2 4 = \log_2 2^2 + \log_2 2^2 = 2 + 2 = 4$

72. $\log_3 3^2 = 2$

    $(\log_3 3)^2 = (1)^2 = 1$

74. $x = \log_a M$ is equivalent to $M = a^x$. Raise both sides of $M = a^x$ to the rth power.

    $M^r = (a^x)^r = a^{rx}$

    Write $M^r = a^{rx}$ in logarithmic form.

    $\log_a M^r = rx$

    Substitute for x.

    $\log_a M^r = r \log_a M$

76. $\log_{10} \dfrac{3}{5} = \log_{10} 3 - \log_{10} 5 \approx 0.4771 - 0.6990 = -0.2219$

78. $\log_{10} \sqrt{5} = \log_{10} 5^{1/2} = \dfrac{1}{2} \log_{10} 5 \approx \dfrac{1}{2}(0.6990) = 0.3495$

Problem Set 11.4, p. 447

2. $\log 7 = 0.8451$

4. $\log 8.1 = 0.9085$

6. $\log 3.49 = 0.5428$

8. $\log 75 = 1.8751$

10. $\log 7500 = 3.8751$

12. $\log 75,000 = 4.8751$

14. $\log 0.371 = -0.4306$

16. $\log 0.0371 = -1.4306$

18. $\log 8,650,000,000$

    $= \log 8.65 \times 10^9$

    $= 9.9370$

20. $\log 0.0000000865$

    $= \log 8.65 \times 10^{-8}$

    $= -7.0630$

22. $x = 6.75$

24. $x = 67.5$

26. $x = 675$

28. $x = 675,000$

30. $x = 537$

32. $x = 3.98 \times 10^{17}$

34. $x = 8.71 \times 10^{86}$ 36. $x = 0.268$

38. $x = 0.0268$ 40. $x = 2.07 \times 10^{-18}$

42.   Because -10 is not in the domain of the logarithmic function.

44. $pH = -\log [H^+]$

$pH = -\log (6.31 \times 10^{-2})$

$= -(\log 6.31 + \log 10^{-2})$

$= -(0.8 + (-2))$

$= -(-1.2)$

$= 1.2$

46. $pH = -\log [H^+]$

$2.1 = -\log [H^+]$

$-2.1 = \log [H^+]$

$[H^+] = \text{antilog} (-2.1) = 7.94 \times 10^{-3}$

48.   For solution A,

$4 = -\log [H^+]$

$-4 = \log [H^+]$

$[H^+] = 10^{-4}.$

For solution B,

$6 = -\log [H^+]$

$-6 = \log [H^+]$

$[H^+] = 10^{-6}.$

Since $10^{-4} = 10^2 \cdot 10^{-6}$, the hydrogen ion concentration of

solution A is $10^2 = 100$ times greater.

Problem Set 11.5, pp. 452–453

2. $3^x = 9$

$3^x = 3^2$

$x = 2$

4. $3^x = 11$

$\log 3^x = \log 11$

$x \log 3 = \log 11$

$x = \dfrac{\log 11}{\log 3}$

$x \approx 2.18$

6. $2^{2x} = 16$

$2^{2x} = 2^4$

$2x = 4$

$x = 2$

8. $2^{2x} = 7$

$\log 2^{2x} = \log 7$

$2x \log 2 = \log 7$

$x = \dfrac{\log 7}{2 \log 2}$

$x \approx 1.40$

10. $9^{x + 1} = 27$

$(3^2)^{x + 1} = 3^3$

$3^{2x + 2} = 3^3$

$2x + 2 = 3$

$2x = 1$

$x = \dfrac{1}{2}$

12. $9^{x + 1} = 14$

$\log 9^{x + 1} = \log 14$

$(x + 1) \log 9 = \log 14$

$x + 1 = \dfrac{\log 14}{\log 9}$

$x = \dfrac{\log 14}{\log 9} - 1$

$x \approx 0.201$

14. $5^{-x} = 625$

$5^{-x} = 5^4$

$-x = 4$

$x = -4$

16. $5^{-x} = 1000$

$\log 5^{-x} = \log 1000$

$-x \log 5 = \log 1000$

$x = - \dfrac{\log 1000}{\log 5}$

$x \approx -4.29$

18. $7^{3x + 1} = 1$

  $7^{3x + 1} = 7^0$

  $3x + 1 = 0$

  $3x = -1$

  $x = -\dfrac{1}{3}$

20. $3^{x^2 - 3x} = \dfrac{1}{9}$

  $3^{x^2 - 3x} = 3^{-2}$

  $x^2 - 3x = -2$

  $x^2 - 3x + 2 = 0$

  $(x - 1)(x - 2) = 0$

  $x = 1$ or $x = 2$

22. $\log_4 7x = \log_4 14$

  $7x = 14$

  $x = 2$

24. $\log_5 (2x - 1) = \log_5 9$

  $2x - 1 = 9$

  $2x = 10$

  $x = 5$

26. $\log (3x + 2) = \log (2x + 5)$

  $3x + 2 = 2x + 5$

  $x = 3$

28. $\log (x + 4) + \log (x - 3) = \log 5x$

  $\log (x + 4)(x - 3) = \log 5x$

  $\log (x^2 + x - 12) = \log 5x$

  $x^2 + x - 12 = 5x$

  $x^2 - 4x - 12 = 0$

  $(x - 6)(x + 2) = 0$

  $x = 6$ or $x = -2$

But $x = -2$ causes $\log (x - 3)$ and $\log 5x$ to be undefined. Therefore the only solution is 6.

30. $\log_3 x = 4$

  $x = 3^4$

  $x = 81$

32. $\log (x + 1) = 2$

  $x + 1 = 10^2$

  $x + 1 = 100$

  $x = 99$

34. $\log_2 (x^2 - 7x) = 3$

   $x^2 - 7x = 2^3$

   $x^2 - 7x - 8 = 0$

   $(x - 8)(x + 1) = 0$

   $x = 8$ or $x = -1$

36. $\log_3 x - \log_3 2 = 1$

   $\log_3 \frac{x}{2} = 1$

   $\frac{x}{2} = 3^1$

   $x = 6$

38. $\log x + \log (x - 21) = 2$

   $\log x(x - 21) = 2$

   $\log (x^2 - 21x) = 2$

   $x^2 - 21x = 10^2$

   $x^2 - 21x - 100 = 0$

   $(x - 25)(x + 4) = 0$

   $x = 25$ or $x = -4$

   But $x = -4$ causes $\log x$ and $\log (x - 21)$ to be undefined. Therefore the only solution is 25.

40. $\log (x + 5) - \log (x - 1) = 1$

   $\log \frac{x + 5}{x - 1} = 1$

   $\frac{x + 5}{x - 1} = 10^1$

   $x + 5 = 10(x - 1)$

   $x + 5 = 10x - 10$

   $15 = 9x$

   $x = \frac{15}{9} = \frac{5}{3}$

42. $V = P(1 + r)^t$

   $1000 = 500(1 + 0.12)^t$

   $1000 = 500(1.12)^t$

   $2 = 1.12^t$

   $\log 2 = \log 1.12^t$

44. $V = P(1 + r)^t$

   $3P = P(1 + 0.20)^t$

   $3P = P(1.2)^t$

   $3 = 1.2^t$

   $\log 3 = \log 1.2^t$

$$\log 2 = t \log 1.12 \qquad\qquad \log 3 = t \log 1.2$$

$$t = \frac{\log 2}{\log 1.12} \qquad\qquad\qquad t = \frac{\log 3}{\log 1.2}$$

$$t \approx 6.1 \text{ yr} \qquad\qquad\qquad t \approx 6 \text{ yr}$$

It will take 6.1 yr for $50,000 to double at 12%.

46.
$$Q = Q_0 \, (2)^{-t/5700}$$

$$0.30 Q_0 = Q_0 \, (2)^{-t/5700}$$

$$0.30 = 2^{-t/5700}$$

$$\log 0.30 = \log 2^{-t/5700}$$

$$\log 0.30 = -\frac{t}{5700} \log 2$$

$$t = -\frac{5700 \log 0.30}{\log 2} \approx 9900 \text{ yr}$$

Problem Set 11.6, pp. 455–456

2. $\log_3 7 = \dfrac{\log_{10} 7}{\log_{10} 3} \approx 1.77$

4. $\log_6 588 = \dfrac{\log_{10} 588}{\log_{10} 6} \approx 3.56$

6. $\log_{11} 3.16 = \dfrac{\log_{10} 3.16}{\log_{10} 11} \approx 0.48$

8. $\log_{1/2} 9.7 = \dfrac{\log_{10} 9.7}{\log_{10} 0.5} \approx -3.28$

10. $\log_2 100 = \dfrac{\log_{10} 100}{\log_{10} 2} \approx 6.64$

12. $\log_3 25 = \dfrac{\log_8 25}{\log_8 3}$

14. $\log_5 11 = \dfrac{\log_{11} 11}{\log_{11} 5} = \dfrac{1}{\log_{11} 5}$

16. $\ln 1480 = 7.2998$ \qquad\qquad 18. $\ln 1000 = 6.9078$

20. $\ln 68 = 4.2195$

22. $\ln 7.42 = 2.0042$

24. $\ln 0.00303 = -5.7992$

26. $\ln e^2 = 2$

28. $\ln e^4 = 4$

30. $\ln \frac{1}{e^2} = \ln e^{-2} = -2$

32. $\ln \sqrt[3]{e} = \ln e^{1/3} = \frac{1}{3}$

34. $e^4 = 54.6$

36. $e^{3.9} = 49.4$

38. $e^{1.26} = 3.53$

40. $e^{-0.8} = 0.449$

42. $V = Pe^{rt}$

$$V = 3000e^{0.12(10)}$$

$$= 3000e^{1.2}$$

$$\approx \$9960.35$$

44. $P = 14.7e^{-0.21a}$

a) $P = 14.7e^{-0.21(5.5)} = 14.7e^{-1.155} \approx 4.6 \ \text{lb/in.}^2$

b) $P = 14.7e^{-0.21(-0.05)} = 14.7e^{0.0105} \approx 14.9 \ \text{lb/in.}^2$

46. By the change-of-base formula,

$$\log_b x = \frac{\log_a x}{\log_a b} \ .$$

Substitute $x = a$.

$$\log_b a = \frac{\log_a a}{\log_a b} = \frac{1}{\log_a b}$$

304

NOTES

CHAPTER 12

SEQUENCES AND SERIES

Problem Set 12.1, pp. 464-466

2. $a_n = 3n + 1$

$a_1 = 3(1) + 1 = 4$

$a_2 = 3(2) + 1 = 7$

$a_3 = 3(3) + 1 = 10$

$a_4 = 3(4) + 1 = 13$

$a_{10} = 3(10) + 1 = 33$

4. $a_n = n^2 - n$

$a_1 = 1^2 - 1 = 0$

$a_2 = 2^2 - 1 = 3$

$a_3 = 3^2 - 1 = 8$

$a_4 = 4^2 - 1 = 15$

$a_{10} = 10^2 - 1 = 99$

6. $a_n = \dfrac{n + 1}{n}$

$a_1 = \dfrac{1 + 1}{1} = 2$

$a_2 = \dfrac{2 + 1}{2} = \dfrac{3}{2}$

$a_3 = \dfrac{3 + 1}{3} = \dfrac{4}{3}$

$a_4 = \dfrac{4 + 1}{4} = \dfrac{5}{4}$

$a_{10} = \dfrac{10 + 1}{10} = \dfrac{11}{10}$

8. $a_n = n - \dfrac{1}{n}$

$a_1 = 1 - \dfrac{1}{1} = 0$

$a_2 = 2 - \dfrac{1}{2} = \dfrac{3}{2}$

$a_3 = 3 - \dfrac{1}{3} = \dfrac{8}{3}$

$a_4 = 4 - \dfrac{1}{4} = \dfrac{15}{4}$

$a_{10} = 10 - \dfrac{1}{10} = \dfrac{99}{10}$

10. $a_n = 8$

$a_1 = 8$

$a_2 = 8$

$a_3 = 8$

$a_4 = 8$

$a_{10} = 8$

12. $a_n = \dfrac{(-1)^n}{n+2}$

$a_1 = \dfrac{(-1)^1}{1+2} = -\dfrac{1}{3}$

$a_2 = \dfrac{(-1)^2}{2+2} = \dfrac{1}{4}$

$a_3 = \dfrac{(-1)^3}{3+2} = -\dfrac{1}{5}$

$a_4 = \dfrac{(-1)^4}{4+2} = \dfrac{1}{6}$

$a_{10} = \dfrac{(-1)^{10}}{10+2} = \dfrac{1}{12}$

14. $a_n = 3^n$

$a_1 = 3^1 = 3$

$a_2 = 3^2 = 9$

$a_3 = 3^3 = 27$

$a_4 = 3^4 = 81$

$a_{10} = 3^{10} = 59{,}049$

16. Each term is three times the term number. Therefore $a_n = 3n$.

18. The difference between successive terms is 2. This suggests $2n$. To obtain the first term 1 when $n = 1$, write $a_n = 2n - 1$.

20. Each term is the fourth power of the term number. Therefore $a_n = n^4$.

22. Each denominator is the cube of the term number. Therefore $a_n = \dfrac{1}{n^3}$.

24. $a_n = (-1)^{n+1}$

26. Each term is three times the term number. This suggests 3n.

Since the signs alternate (beginning with a negative),

$a_n = (-1)^n 3n.$

28. $\displaystyle\sum_{n=1}^{7} 10n = 10 \cdot 1 + 10 \cdot 2 + 10 \cdot 3 + 10 \cdot 4 + 10 \cdot 5 + 10 \cdot 6 + 10 \cdot 7$

$= 10 + 20 + 30 + 40 + 50 + 60 + 70$

$= 280$

30. $\displaystyle\sum_{n=1}^{7} 10 = 10 + 10 + 10 + 10 + 10 + 10 + 10$

$= 70$

32. $\displaystyle\sum_{n=2}^{6} (2n - 1) = (2 \cdot 2 - 1) + (2 \cdot 3 - 1) + (2 \cdot 4 - 1) + (2 \cdot 5 - 1) + (2 \cdot 6 - 1)$

$= 3 + 5 + 7 + 9 + 11$

$= 35$

34. $\displaystyle\sum_{n=2}^{6} 2n - 1 = (\sum_{n=2}^{6} 2n) - 1$

$= (2 \cdot 2 + 2 \cdot 3 + 2 \cdot 4 + 2 \cdot 5 + 2 \cdot 6) - 1$

$= (4 + 6 + 8 + 10 + 12) - 1$

$= 39$

36. $\displaystyle\sum_{n=0}^{4} (n^2 - 4n + 5)$

$= (0^2 - 4 \cdot 0 + 5) + (1^2 - 4 \cdot 1 + 5) + (2^2 - 4 \cdot 2 + 5)$

$+ (3^2 - 4 \cdot 3 + 5) + (4^2 - 4 \cdot 4 + 5)$

$= 5 + 2 + 1 + 2 + 5$

$= 15$

38. $\displaystyle\sum_{n=1}^{3}(n^3 + n) = (1^3 + 1) + (2^3 + 2) + (3^3 + 3)$

$= 2 + 10 + 30$

$= 42$

40. $\displaystyle\sum_{n=0}^{3}\frac{1}{n+1} = \frac{1}{0+1} + \frac{1}{1+1} + \frac{1}{2+1} + \frac{1}{3+1}$

$= 1 + \frac{1}{2} + \frac{1}{3} + \frac{1}{4}$

$= \frac{25}{12}$

42. $\displaystyle\sum_{n=0}^{3}(-n)^{n+1} = (-0)^{0+1} + (-1)^{1+1} + (-2)^{2+1} + (-3)^{3+1}$

$= 0^1 + (-1)^2 + (-2)^3 + (-3)^4$

$= 0 + 1 + (-8) + 81$

$= 74$

44. $\displaystyle\sum_{n=1}^{5}\frac{x^{2n}}{2n} = \frac{x^{2\cdot1}}{2\cdot1} + \frac{x^{2\cdot2}}{2\cdot2} + \frac{x^{2\cdot3}}{2\cdot3} + \frac{x^{2\cdot4}}{2\cdot4} + \frac{x^{2\cdot5}}{2\cdot5}$

$= \frac{x^2}{2} + \frac{x^4}{4} + \frac{x^6}{6} + \frac{x^8}{8} + \frac{x^{10}}{10}$

46. $1 + 5 + 9 + 13 + 17 + 21 = \displaystyle\sum_{n=0}^{5}(4n + 1)$

48. $5 + 9 + 13 + 17 + 21 + 25 = \displaystyle\sum_{n=1}^{6}(4n + 1)$

50. $2 + 4 + 6 + 8 = \displaystyle\sum_{n=1}^{4}2n$

52. $2 + 4 + 8 + 16 = \displaystyle\sum_{n=1}^{4}2^n$

54. $\dfrac{3}{4} + \dfrac{4}{5} + \dfrac{5}{6} + \dfrac{6}{7} = \displaystyle\sum_{n=3}^{6}\frac{n}{n+1}$

56. $0 + 3 + 8 + 15 + 24 = \sum_{n=1}^{5} (n^2 - 1)$

58. $x + x^2 + x^3 + x^4 + x^5 = \sum_{n=1}^{5} x^n$

60. $\sum_{n=1}^{5} n^2 = 1^2 + 2^2 + 3^2 + 4^2 + 5^2 = 55$

   $\sum_{k=1}^{5} k^2 = 1^2 + 2^2 + 3^2 + 4^2 + 5^2 = 55$

62. First year's salary = $10,000

   Second year's salary = 10,000 + 10,000(0.10)

   = 10,000(1 + 0.10)

   = 10,000(1.1)

   = $11,000

   Third year's salary = 11,000 + 11,000(0.10)

   = 11,000(1 + 0.10)

   = 11,000(1.1)

   = $12,100

   Fourth year's salary = 12,100(1.1)

   = $13,310

   Fifth year's salary = 13,310(1.1)

   = $14,641

64. Each term after the second is the sum of the preceding two terms.

66. $2^0 + 2^1 + 2^2 + 2^3 + 2^4 + \ldots + 2^{63} = \sum_{n=0}^{63} 2^n$

   a) $2^0 + 2^1 + 2^2 + 2^3 + 2^4 + 2^5 + 2^6 + 2^7 + 2^8 + 2^9$

   = 1023 grains

   b) $2^{63} \approx 9.2 \times 10^{18}$ grains

Problem Set 12.2, pp. 470-471

2. Arithmetic; $d = 3 - 1 = 2$; 11, 13

4. Arithmetic; $d = 11 - 4 = 7$; 39, 46

6. Arithmetic; $d = 12 - 4 = 8$; 28, 36

8. Arithmetic; $d = 3 - 9 = -6$; -21, -27

10. Arithmetic; $d = \frac{5}{3} - \frac{4}{3} = \frac{1}{3}$; $\frac{8}{3}$, 3

12. Nonarithmetic

14. $a_n = a_1 + (n - 1)d$

   $a_{21} = 2 + (21 - 1)5$

   $\quad\; = 2 + (20)5$

   $\quad\; = 102$

16. $a_n = a_1 + (n - 1)d$

   $a_{111} = 2 + (111 - 1)5$

   $\quad\;\; = 2 + (110)5$

   $\quad\;\; = 552$

18. $a_n = a_1 + (n - 1)d$

   $a_n = 2 + (n - 1)5$

   $a_n = 2 + 5n - 5$

   $a_n = 5n - 3$

20. $d = 11 - 15 = -4$

   $a_n = a_1 + (n - 1)d$

   $a_{25} = 15 + (25 - 1)(-4)$

   $\quad\;\; = 15 + (24)(-4)$

   $\quad\;\; = -81$

22. $d = 3 - \frac{1}{2} = \frac{5}{2}$

   $a_n = a_1 + (n - 1)d$

   $a_{99} = \frac{1}{2} + (99 - 1)\frac{5}{2}$

   $\quad\; = \frac{1}{2} + (98)\frac{5}{2}$

   $\quad\; = \frac{491}{2}$

24. $a_n = a_1 + (n - 1)d$

   $a_6 = 7 + (6 - 1)d$

   $a_6 = 7 + 5d$

   $22 = 7 + 5d$

   $15 = 5d$

   $3 = d$

26. $a_n = a_1 + (n - 1)d$

    $a_8 = -2 + (8 - 1)d$

    $a_8 = -2 + 7d$

    $19 = -2 + 7d$

    $21 = 7d$

    $3 = d$

28. $S_n = \frac{n}{2}(a_1 + a_n)$

    $S_{10} = \frac{10}{2}(a_1 + a_{10})$

    $= 5(6 + 15)$

    $= 5(21)$

    $= 105$

30. $a_n = 2n + 3$

    $a_1 = 2(1) + 3 = 5$

    $a_{20} = 2(20) + 3 = 43$

    $S_n = \frac{n}{2}(a_1 + a_n)$

    $S_{20} = \frac{20}{2}(a_1 + a_{20})$

    $= 10(5 + 43)$

    $= 10(48)$

    $= 480$

32. $a_n = a_1 + (n - 1)d$

    $a_{10} = \frac{1}{2} + (10 - 1)\frac{1}{3}$

    $= \frac{1}{2} + 3$

    $= \frac{7}{2}$

    $S_n = \frac{n}{2}(a_1 + a_n)$

    $S_{10} = \frac{10}{2}(a_1 + a_{10})$

    $= 5(\frac{1}{2} + \frac{7}{2})$

    $= 5(\frac{8}{2})$

    $= 20$

34. $d = 11 - 2 = 9$

    $a_n = a_1 + (n - 1)d$

    $a_{19} = 2 + (19 - 1)9$

    $= 2 + (18)9$

    $= 164$

36. $\sum\limits_{k=1}^{2000} (6k - 1)$

    $= 5 + 11 + 17 + \ldots + 11,999$

    $S_n = \frac{n}{2}(a_1 + a_n)$

    $S_{2000} = \frac{2000}{2}(a_1 + a_{2000})$

    $= 1000(5 + 11,999)$

$$S_n = \frac{n}{2}(a_1 + a_n)$$

$$S_{19} = \frac{19}{2}(a_1 + a_{19})$$

$$= \frac{19}{2}(2 + 164)$$

$$= 1577$$

$$= 1000(12,004)$$

$$= 12,004,000$$

38. $d = 10 - 6 = 4$

$a_n = a_1 + (n - 1)d$

$58 = 6 + (n - 1)4$

$58 = 6 + 4n - 4$

$58 = 4n + 2$

$56 = 4n$

$14 = n$

40. $1 + 2 + 3 + \ldots + 200$

$$S_n = \frac{n}{2}(a_1 + a_n)$$

$$S_{200} = \frac{200}{2}(a_1 + a_{200})$$

$$= 100(1 + 200)$$

$$= 100(201)$$

$$= 20,100$$

42. $50 + 60 + 70 + \ldots$

$d = 60 - 50 = 10$

$a_n = a_1 + (n - 1)d$

$a_{21} = 50 + (21 - 1)10$

$\quad = 50 + (20)10$

$\quad = 250$

$$S_n = \frac{n}{2}(a_1 + a_n)$$

$$S_{21} = \frac{21}{2}(a_1 + a_{21})$$

$$= \frac{21}{2}(50 + 250)$$

$$= \frac{21}{2}(300)$$

$$= \$3150$$

44. $97, 94, 91, \ldots$

$d = 94 - 97 = -3$

a) $a_n = a_1 + (n - 1)d$

$a_{23} = 97 + (23 - 1)(-3)$

$\quad = 97 + (22)(-3)$

$\quad = 31$ bottles

b) $S_n = \frac{n}{2}(a_1 + a_n)$

$S_{23} = \frac{23}{2}(a_1 + a_{23})$

$\quad = \frac{23}{2}(97 + 31)$

$\quad = \frac{23}{2}(128)$

$\quad = 1472$ bottles

46. $S_n = \frac{n}{2}(a_1 + a_n)$

Replace a_n by $a_1 + (n - 1)d$.

$S_n = \frac{n}{2}[a_1 + a_1 + (n - 1)d]$

$S_n = \frac{n}{2}[2a_1 + (n - 1)d]$

48. $1 + 2 + 3 + \ldots$

$d = 2 - 1 = 1$

18 yr = 18(365) days = 6570 days

$a_n = a_1 + (n - 1)d$

$a_{6570} = 1 + (6570 - 1)1 = 1 + 6569 = 6570$

$S_n = \frac{n}{2}(a_1 + a_n)$

$S_{6570} = \frac{6570}{2}(a_1 + a_{6570})$

$S_{6570} = \frac{6570}{2}(1 + 6570) = 21,585,735¢ = \$215,857.35$

Problem Set 12.3, pp. 475–476

2.   Geometric; r = 3; 243, 729

4.   Geometric; $r = \frac{1}{4}$; $\frac{1}{256}$, $\frac{1}{1024}$

6.   Nongeometric

8.   Geometric; r = 2; −16, −32

10.   Nongeometric

12.   Geometric; r = −1; −7, 7

14.   Geometric; $r = -\frac{1}{3}$; $-\frac{1}{27}$, $\frac{1}{81}$

16. $a_n = a_1 r^{n-1}$

$a_7 = 3(2)^{7-1}$

$= 3(2)^6$

$= 3(64)$

$= 192$

18. $a_n = a_1 r^{n-1}$

$a_n = 3(2)^{n-1}$

20. $r = \frac{3}{9} = \frac{1}{3}$

$a_n = a_1 r^{n-1}$

$a_5 = 9(\frac{1}{3})^{5-1}$

$= 9(\frac{1}{3})^4$

$= 9(\frac{1}{81})$

$= \frac{1}{9}$

22. $r = -\frac{1}{3} \div \frac{1}{2} = -\frac{1}{3} \cdot \frac{2}{1} = -\frac{2}{3}$

$a_n = a_1 r^{n-1}$

$a_5 = \frac{1}{2}(-\frac{2}{3})^{5-1}$

$= \frac{1}{2}(-\frac{2}{3})^4$

$= \frac{1}{2}(\frac{16}{81})$

$= \frac{8}{81}$

24. $r = \frac{-8}{-4} = 2$

$a_n = a_1 r^{n-1}$

$a_4 = -4(2)^{4-1}$

$= -4(2)^3$

$= -4(8)$

$= -32$

26. $a_n = a_1 r^{n-1}$

$a_4 = a_1 r^{4-1}$

$a_4 = a_1 r^3$

$0.32 = 5r^3$

$0.064 = r^3$

$\sqrt[3]{0.064} = r$

$0.4 = r$

28. $r = \dfrac{1}{1} = 1$

$a_n = a_1 r^{n-1}$

$a_{67} = 1(1)^{67-1}$

$= 1(1)^{66}$

$= 1(1)$

$= 1$

30. $r = \dfrac{1}{-1} = -1$

$a_n = a_1 r^{n-1}$

$a_n = -1(-1)^{n-1}$

$a_n = (-1)^1 (-1)^{n-1}$

$a_n = (-1)^{1+n-1}$

$a_n = (-1)^n$

32. $S_n = \dfrac{a_1(1-r^n)}{1-r}$

$S_4 = \dfrac{9(1-5^4)}{1-5}$

$= \dfrac{9(1-625)}{-4}$

$= \dfrac{9(-624)}{-4}$

$= 1404$

34. $S_n = \dfrac{a_1(1-r^n)}{1-r}$

$S_7 = \dfrac{80[1-(\frac{1}{2})^7]}{1-\frac{1}{2}}$

$= \dfrac{80[1-\frac{1}{128}]}{\frac{1}{2}}$

$= \dfrac{80[\frac{127}{128}]}{\frac{1}{2}}$

$= \dfrac{80}{1} \cdot \dfrac{127}{128} \cdot \dfrac{2}{1}$

$= \dfrac{635}{4}$

36. $r = \dfrac{-6}{3} = -2$

$S_n = \dfrac{a_1(1-r^n)}{1-r}$

$S_9 = \dfrac{3[1-(-2)^9]}{1-(-2)}$

$= \dfrac{3[1-(-512)]}{3}$

$= 513$

38. $r = \dfrac{1}{-1} = -1$

$S_n = \dfrac{a_1(1-r^n)}{1-r}$

$S_{100} = \dfrac{-1[1-(-1)^{100}]}{1-(-1)}$

$= \dfrac{-1[1-1]}{2}$

$= \dfrac{-1[0]}{2}$

$= 0$

$$S_{101} = \frac{-1[1 - (-1)^{101}]}{1 - (-1)}$$

$$= \frac{-1[1 - (-1)]}{2}$$

$$= \frac{-1[2]}{2}$$

$$= -1$$

40. $\displaystyle\sum_{k=1}^{10} 81\left(\frac{2}{3}\right)^k = 81\left(\frac{2}{3}\right) + 81\left(\frac{2}{3}\right)^2 + 81\left(\frac{2}{3}\right)^3 + \ldots + 81\left(\frac{2}{3}\right)^{10}$

$$r = \frac{81(2/3)^2}{81(2/3)} = \frac{2}{3}$$

$$S_n = \frac{a_1(1 - r^n)}{1 - r}$$

$$S_{10} = \frac{81\left(\frac{2}{3}\right)[1 - \left(\frac{2}{3}\right)^{10}]}{1 - \frac{2}{3}} = \frac{54\left[1 - \frac{1024}{59,049}\right]}{\frac{1}{3}}$$

$$= \frac{18\left[\frac{59,049 - 1024}{59,049}\right]}{\frac{1}{3}}$$

$$= \frac{54}{1} \cdot \frac{58,025}{59,049} \cdot \frac{3}{1}$$

$$= \frac{116,050}{729}$$

42. $a_1 = 80\left(\frac{3}{4}\right) = 60$

$$a_n = a_1 r^{n-1}$$

$$a_4 = 60\left(\frac{3}{4}\right)^{4-1}$$

$$= 60\left(\frac{3}{4}\right)^3$$

$$= 60\left(\frac{27}{64}\right)$$

$$= \frac{405}{16}$$

$$= 25\frac{5}{16} \text{ ft}$$

44. $S_n = \dfrac{a_1(1 - r^n)}{1 - r}$

$$S_n = \frac{a_1 - a_1 r^n}{1 - r}$$

$$S_n = \frac{a_1 - a_1 r^{n-1} \cdot r}{1 - r}$$

Replace $a_1 r^{n-1}$ with a_n.

$$S_n = \frac{a_1 - a_n r}{1 - r}$$

46.   Cost in 1 yr = 12,000 + 12,000(0.09)

$$= 12,000(1 + 0.09)$$

$$= 12,000(1.09)$$

Cost in 2 yr = 12,000(1.09) + 12,000(1.09)(0.09)

$$= 12,000(1.09)(1 + 0.09)$$

$$= 12,000(1.09)(1.09)$$

$$= 12,000(1.09)^2$$

Therefore the yearly costs form the geometric sequence

$$12,000(1.09),\ 12,000(1.09)^2,\ 12,000(1.09)^3,\ \ldots\ .$$

$$a_n = a_1 r^{n-1}$$

$$a_{20} = 12,000(1.09)(1.09)^{20-1}$$

$$a_{20} = 13,080(1.09)^{19}$$

$$a_{20} \approx \$67,252.93$$

Problem Set 12.4, pp. 478-479

2. $S = \dfrac{a_1}{1-r}$

$$S = \dfrac{12}{1 - \dfrac{2}{3}}$$

$$= \dfrac{12}{1/3}$$

$$= \dfrac{12}{1} \cdot \dfrac{3}{1} = 36$$

4.   Sum does not exist since

$$|r| \geq 1.$$

6. $S = \dfrac{a_1}{1-r}$

$$S = \dfrac{3}{1 - (-\tfrac{1}{2})}$$

8. $r = \dfrac{1}{3} \div 1 = \dfrac{1}{3}$

$$S = \dfrac{a_1}{1-r}$$

$$S = \dfrac{1}{1 - \dfrac{1}{3}}$$

$$= \frac{3}{1 + \frac{1}{2}}$$

$$= \frac{3}{3/2}$$

$$= \frac{3}{1} \cdot \frac{2}{3}$$

$$= 2$$

$$= \frac{1}{2/3}$$

$$= \frac{3}{2}$$

10. $r = \frac{7}{10} = 0.7$

$$S = \frac{a_1}{1 - r}$$

$$S = \frac{10}{1 - 0.7}$$

$$= \frac{10}{0.3}$$

$$= \frac{100}{3}$$

12. $r = \frac{-20}{25} = -\frac{4}{5}$

$$S = \frac{a_1}{1 - r}$$

$$S = \frac{25}{1 - (-\frac{4}{5})}$$

$$= \frac{25}{1 + \frac{4}{5}}$$

$$= \frac{25}{9/5}$$

$$= \frac{25}{1} \cdot \frac{5}{9}$$

$$= \frac{125}{9}$$

14. $r = -\frac{1}{12} \div \frac{1}{16} = -\frac{1}{12} \cdot \frac{16}{1} = -\frac{4}{3}$

Sum does not exist since $|r| \geq 1$.

16. $r = \frac{-1}{1} = -1$

Sum does not exist since $|r| \geq 1$.

18. $S = \dfrac{a_1}{1 - r}$

$18 = \dfrac{a_1}{1 - \dfrac{1}{6}}$

$18 = \dfrac{a_1}{5/6}$

$\dfrac{5}{6} \cdot 18 = \dfrac{5}{6} \cdot \dfrac{a_1}{5/6}$

$15 = a_1$

20. $S = \dfrac{a_1}{1 - r}$

$21 = \dfrac{6}{1 - r}$

$21(1 - r) = 6$

$21 - 21r = 6$

$-21r = -15$

$r = \dfrac{-15}{-21}$

$r = \dfrac{5}{7}$

22. $0.\overline{6} = 0.6 + 0.06 + 0.006 + \ldots$

$S = \dfrac{a_1}{1 - r} = \dfrac{0.6}{1 - 0.1} = \dfrac{0.6}{0.9} = \dfrac{6}{9} = \dfrac{2}{3}$

24. $0.\overline{18} = 0.18 + 0.0018 + 0.000018 + \ldots$

$S = \dfrac{a_1}{1 - r} = \dfrac{0.18}{1 - 0.01} = \dfrac{0.18}{0.99} = \dfrac{18}{99} = \dfrac{2}{11}$

26. $0.\overline{08} = 0.08 + 0.0008 + 0.000008 + \ldots$

$S = \dfrac{a_1}{1 - r} = \dfrac{0.08}{1 - 0.01} = \dfrac{0.08}{0.99} = \dfrac{8}{99}$

28. $0.\overline{471} = 0.471 + 0.000471 + 0.000000471 + \ldots$

$S = \dfrac{a_1}{1 - r} = \dfrac{0.471}{1 - 0.001} = \dfrac{0.471}{0.999} = \dfrac{471}{999} = \dfrac{157}{333}$

30. $9.\overline{9} = 9 + 0.9 + 0.09 + 0.009 + \ldots$

$S = \dfrac{a_1}{1 - r} = \dfrac{9}{1 - 0.1} = \dfrac{9}{0.9} = \dfrac{90}{9} = 10$

32. $0.\overline{8} = 0.8 + 0.08 + 0.008 + \ldots$

$S = \dfrac{a_1}{1 - r} = \dfrac{0.8}{1 - 0.1} = \dfrac{0.8}{0.9} = \dfrac{8}{9}$

Therefore $7.\overline{8} = 7\dfrac{8}{9} = \dfrac{71}{9}.$

34. $\displaystyle\sum_{n=1}^{\infty} 5\left(\frac{3}{4}\right)^n = 5\left(\frac{3}{4}\right) + 5\left(\frac{3}{4}\right)^2 + 5\left(\frac{3}{4}\right)^3 + \dots$

$S = \dfrac{a_1}{1-r} = \dfrac{5\left(\frac{3}{4}\right)}{1-\frac{3}{4}} = \dfrac{15/4}{1/4} = \dfrac{15}{4}\cdot\dfrac{4}{1} = 15$

36. $45 + 45\left(\frac{4}{5}\right) + 45\left(\frac{4}{5}\right)^2 + \dots$

$S = \dfrac{a_1}{1-r} = \dfrac{45}{1-\frac{4}{5}}$

$= \dfrac{45}{1/5}$

$= \dfrac{45}{1}\cdot\dfrac{5}{1}$

$= 225 \text{ cm}$

38. $S = \dfrac{a_1}{1-r}$

$S = \dfrac{1}{1-\frac{1}{2}} = \dfrac{1}{1/2} = \dfrac{1}{1}\cdot\dfrac{2}{1} = 2$

$S_n = \dfrac{a_1(1-r^n)}{1-r}$

$S_5 = \dfrac{1[1-\left(\frac{1}{2}\right)^5]}{1-\frac{1}{2}} = \dfrac{1-\frac{1}{32}}{\frac{1}{2}} = \dfrac{\frac{31}{32}}{\frac{1}{2}} = \dfrac{31}{32}\cdot\dfrac{2}{1} = 1.9375$

$S_{10} = \dfrac{1[1-\left(\frac{1}{2}\right)^{10}]}{1-\frac{1}{2}} = \dfrac{1-\frac{1}{1024}}{\frac{1}{2}} = \dfrac{\frac{1023}{1024}}{\frac{1}{2}} = \dfrac{1023}{1024}\cdot\dfrac{2}{1} \approx 1.9980$

$S_{15} = \dfrac{1[1-\left(\frac{1}{2}\right)^{15}]}{1-\frac{1}{2}} = \dfrac{1-\frac{1}{32,768}}{\frac{1}{2}} = \dfrac{\frac{32,767}{32,768}}{\frac{1}{2}} \approx 1.9999$

Problem Set 12.5, pp. 483-484

2.

$$
\begin{array}{ccccccccccccccccc}
&&&&&&&& 1 \\
&&&&&&& 1 && 1 \\
&&&&&& 1 && 2 && 1 \\
&&&&& 1 && 3 && 3 && 1 \\
&&&& 1 && 4 && 6 && 4 && 1 \\
&&& 1 && 5 && 10 && 10 && 5 && 1 \\
&& 1 && 6 && 15 && 20 && 15 && 6 && 1 \\
& 1 && 7 && 21 && 35 && 35 && 21 && 7 && 1 \\
1 && 8 && 28 && 56 && 70 && 56 && 28 && 8 && 1
\end{array}
$$

$$(a + b)^8 = a^8 + 8a^7b + 28a^6b^2 + 56a^5b^3 + 70a^4b^4 + 56a^3b^5$$

$$+ 28a^2b^6 + 8ab^7 + b^8$$

4. $\binom{7}{2} = \dfrac{7!}{2!\,(7-2)!} = \dfrac{7\cdot6\cdot5!}{2\cdot1\cdot5!} = 21$

6. $\binom{7}{5} = \dfrac{7!}{5!\,(7-5)} = \dfrac{7\cdot6\cdot5!}{5!2!} = \dfrac{7\cdot6\cdot5!}{5!\cdot2\cdot1} = 21$

8. $\binom{10}{7} = \dfrac{10!}{7!\,(10-7)!} = \dfrac{10\cdot9\cdot8\cdot7!}{7!3!} = \dfrac{10\cdot9\cdot8\cdot7!}{7!\cdot3\cdot2\cdot1} = 120$

10. $\binom{9}{1} = \dfrac{9!}{1!\,(9-1)!} = \dfrac{9\cdot8!}{1\cdot8!} = 9$

12. $\binom{9}{0} = \dfrac{9!}{0!\,(9-0)!} = \dfrac{9!}{1\cdot9!} = 1$

14. $\binom{20}{19} = \dfrac{20!}{19!\,(20-19)!} = \dfrac{20\cdot19!}{19!\cdot1!} = 20$

16. $(x + y)^5 = \binom{5}{0}x^5 + \binom{5}{1}x^4y + \binom{5}{2}x^3y^2 + \binom{5}{3}x^2y^3 + \binom{5}{4}xy^4 + \binom{5}{5}y^5$

$$= 1x^5 + 5x^4y + 10x^3y^2 + 10x^2y^3 + 5xy^4 + 1y^5$$

$$= x^5 + 5x^4y + 10x^3y^2 + 10x^2y^3 + 5xy^4 + y^5$$

18. $(x + 5)^4 = \binom{4}{0}x^4 + \binom{4}{1}x^3(5) + \binom{4}{2}x^2(5)^2 + \binom{4}{3}x(5)^3 + \binom{4}{4}(5)^4$

$$= 1x^4 + 4x^3(5) + 6x^2(25) + 4x(125) + 1(625)$$

$$= x^4 + 20x^3 + 150x^2 + 500x + 625$$

20. $(x - y)^3$

$= [x + (-y)]^3$

$= \binom{3}{0}x^3 + \binom{3}{1}x^2(-y) + \binom{3}{2}x(-y)^2 + \binom{3}{3}(-y)^3$

$= 1x^3 + 3x^2(-y) + 3xy^2 + 1(-y^3)$

$= x^3 - 3x^2y + 3xy^2 - y^3$

22. $(3x + y)^4$

$= [(3x) + y]^4$

$= \binom{4}{0}(3x)^4 + \binom{4}{1}(3x)^3y + \binom{4}{2}(3x)^2y^2 + \binom{4}{3}(3x)y^3 + \binom{4}{4}y^4$

$= 1(81x^4) + 4(27x^3)y + 6(9x^2)y^2 + 4(3x)y^3 + 1y^4$

$= 81x^4 + 108x^3y + 54x^2y^2 + 12xy^3 + y^4$

24. $(a - 2)^5$

$= [a + (-2)]^5$

$= \binom{5}{0}a^5 + \binom{5}{1}a^4(-2) + \binom{5}{2}a^3(-2)^2 + \binom{5}{3}a^2(-2)^3 + \binom{5}{4}a(-2)^4 + \binom{5}{5}(-2)^5$

$= 1a^5 + 5a^4(-2) + 10a^3(4) + 10a^2(-8) + 5a(16) + 1(-32)$

$= a^5 - 10a^4 + 40a^3 - 80a^2 + 80a - 32$

26. $(3m - 4)^5$

$= [(3m) + (-4)]^5$

$= \binom{5}{0}(3m)^5 + \binom{5}{1}(3m)^4(-4) + \binom{5}{2}(3m)^3(-4)^2 + \binom{5}{3}(3m)^2(-4)^3$

$\qquad\qquad\qquad + \binom{5}{4}(3m)(-4)^4 + \binom{5}{5}(-4)^5$

$= 1(243m^5) + 5(81m^4)(-4) + 10(27m^3)(16) + 10(9m^2)(-64)$

$\qquad\qquad\qquad + 5(3m)(256) + 1(-1024)$

$= 243m^5 - 1620m^4 + 4320m^3 - 5760m^2 + 3840m - 1024$

28. $(x + \frac{y}{3})^4$

$$= \binom{4}{0}x^4 + \binom{4}{1}x^3(\frac{y}{3}) + \binom{4}{2}x^2(\frac{y}{3})^2 + \binom{4}{3}x(\frac{y}{3})^3 + \binom{4}{4}(\frac{y}{3})^4$$

$$= 1x^4 + 4x^3(\frac{y}{3}) + 6x^2(\frac{y^2}{9}) + 4x(\frac{y^3}{27}) + 1(\frac{y^4}{81})$$

$$= x^4 + \frac{4}{3}x^3y + \frac{2}{3}x^2y^2 + \frac{4}{27}xy^3 + \frac{1}{81}y^4$$

30. $(t^2 - 1)^6$

$$= [(t^2) + (-1)]^6$$

$$= \binom{6}{0}(t^2)^6 + \binom{6}{1}(t^2)^5(-1) + \binom{6}{2}(t^2)^4(-1)^2 + \binom{6}{3}(t^2)^3(-1)^3$$

$$+ \binom{6}{4}(t^2)^2(-1)^4 + \binom{6}{5}(t^2)(-1)^5 + \binom{6}{6}(-1)^6$$

$$= t^{12} - 6t^{10} + 15t^8 - 20t^6 + 15t^4 - 6t^2 + 1$$

32. The first three terms of $(a + b)^{30}$ are

$$\binom{30}{0}a^{30} + \binom{30}{1}a^{29}b + \binom{30}{2}a^{28}b^2$$

$$= a^{30} + 30a^{29}b + 435a^{28}b^2.$$

34. The first three terms of $(x + 10)^{25}$ are

$$\binom{25}{0}x^{25} + \binom{25}{1}x^{24}(10) + \binom{25}{2}x^{23}(10)^2$$

$$= x^{25} + 25x^{24}(10) + 300x^{23}(100)$$

$$= x^{25} + 250x^{24} + 30,000x^{23}.$$

36. The first three terms of $(t^2 + 1)^{13}$ are

$$\binom{13}{0}(t^2)^{13} + \binom{13}{1}(t^2)^{12}(1) + \binom{13}{2}(t^2)^{11}(1)^2$$

$$= t^{26} + 13t^{24}(1) + 78t^{22}(1)$$

$$= t^{26} + 13t^{24} + 78t^{22}.$$

38. $(a + b)^{11}$

rth term $= \binom{n}{r-1} a^{n - (r - 1)} b^{r - 1}$

5th term $= \binom{11}{4} a^{11 - 4} b^4$

$\qquad = \frac{11!}{4!7!} a^7 b^4$

$\qquad = 330 a^7 b^4$

40. $(p - 2q)^{16} = [p + (-2q)]^{16}$

rth term $= \binom{n}{r-1} a^{n - (r - 1)} b^{r - 1}$

4th term $= \binom{16}{3} p^{16 - 3} (-2q)^3$

$\qquad = \frac{16!}{3!13!} p^{13} (-8q^3)$

$\qquad = 560 p^{13} (-8q^3)$

$\qquad = -4480 p^{13} q^3$

42. $(x + \frac{y}{3})^{10}$, middle term = 6th term

rth term $= \binom{n}{r-1} a^{n - (r - 1)} b^{r - 1}$

6th term $= \binom{10}{5} x^{10 - 5} (\frac{y}{3})^5$

$\qquad = \frac{10!}{5!5!} x^5 (\frac{y^5}{243})$

$\qquad = 252 x^5 \frac{y^5}{243}$

$\qquad = \frac{28}{27} x^5 y^5$

44. a) $0! = 1$

   b) $7! = 5040$

   c) $11! = 39,916,800$

   d) $69! \approx 1.71 \times 10^{98}$

   e) $70!$ is too large for your calculator to display because it is greater than 10^{100}.